BLUEPRINT
FOR A
GREEN PLANET

BLUEPRINT
FOR A
GREEN PLANET

Your practical guide to restoring the world's environment

JOHN SEYMOUR
HERBERT GIRARDET

Illustrated by IAN PENNEY

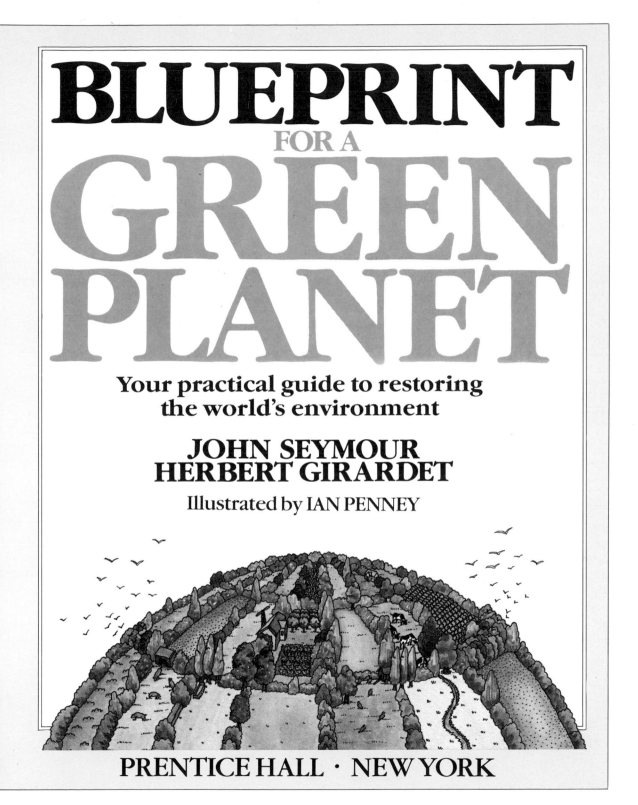

PRENTICE HALL · NEW YORK

Project Editor David Burnie
Art Editor Jane Owen
Editor Rebecca Abrams
Picture Researcher Diana Korchien

Published in 1987 by Prentice Hall Press
A Division of Simon & Schuster, Inc.
Gulf + Western Building
One Gulf + Western Plaza
New York, NY 10023

Originally published in 1987 by
Dorling Kindersley Limited, London

PRENTICE HALL PRESS is a trademark of Simon & Schuster, Inc.

Library of Congress Cataloguing-in-Publication Data

Seymour, John, 1914–
    Blueprint for a green planet.
    Includes index.
    1. Pollution—Environmental aspects.
2. Environmental protection.   I. Girardet,
Herbert.   II. Title.
QH545.A1S49 1987   363.7′3   86-25571
ISBN 0-13-079625-5
ISBN 0-13-079609-3 (pbk.)

Manufactured in Italy

10   9   8   7   6   5   4   3   2   1

First Prentice Hall Edition

CONTENTS

FOREWORD

Our planet is under the most severe strain. Deserts are advancing wherever deserts are. Forests are being destroyed at an alarming rate wherever forests are. Lakes, rivers, and even seas are being degraded and polluted. Perhaps most sinister of all, the very air we have to breathe is becoming dramatically altered for the worse. And air, like disease, knows no frontiers.

Living in green northern Europe, or the more temperate parts of North America, it is easy to be unaware of this. But people over by far the greater part of our planet, if they have any eyes, are very much aware of it.

A number of books, on both sides of the Atlantic, have called attention to the dangers and have tried to alert people to what is happening. But what has been lacking is a single book telling people what to *do* about the ills that afflict our world. It has been necessary to cry "doom!" It has been extremely important to say to us all: "Look, we and succeeding generations have to live on this planet! We ruin it at our peril!" But too much talk of impending doom may work against us. We may feel that the deterioration of our planet is inevitable and that we, as individuals, can do nothing to stave it off. We may feel powerless. *Après nous la déluge.*

TURNING THE TIDE

The authors of this book recently spent over two years traveling throughout Europe, America, Asia, and Africa for a series of BBC films. What we witnessed was not calculated to cheer us up: we saw, among other things, deforestation, pollution, and the very ground beneath our feet, the soil on which we depend, wasting away.

But instead of remaining thoroughly depressed, we came to a different conclusion. It is this. There *are* things that the ordinary person can do. The world's decline is not inevitable. We are *not* powerless.

We are members of an intelligent species. We have been given foresight and the instinct of self-preservation – preservation not only for ourselves but for our children and our children's children too. We *can* prevent the deluge. If each of us – each individual – simply becomes aware of the dangers and does what he or she can to avoid them, then as a species, we can continue to inhabit this planet. Far from turning our world into a wasteland, we can turn it into a paradise again.

A PLANET FIT FOR THE FUTURE

This may be the only planet in the universe with life on it. More probably there are millions of others, if not billions. But we humans are the only self-conscious beings that we know of. We are enormously privileged to be intelligently alive. But, having this privilege, we also carry a responsibility to use our intelligence to beneficial ends.

Some of us may believe that we are here for a purpose. And this purpose is not just to seek ephemeral pleasures and diversions, but to play a real part in the ordering of this planet so as to make it a decent and beautiful place, not only for ourselves, but for all other forms of life too.

But whether we think along these lines or not, our common sense must tell us that we have a bounden duty to hand down this planet to our descendants as good as we found it or better – not worse.

Whether or not we are rewarded or punished in an afterlife we can never know in this one. But one thing is certain, we can punish ourselves *during* our time on Earth. What we should be doing instead is playing our proper part in the march of life on this planet by leaving it fit for the generations to come.

WHY INDIVIDUAL ACTION MATTERS

An individual can have no control over the thoughts or actions of others, but he or she should have complete control over his or her own actions. Each one of us can (and should) try to influence other people – influence governments even – to prevent the devastation of our environment. But we have complete responsibility for what we do, or fail to do, ourselves, and we should be held accountable for what we do, or fail to do, during this particular life on this particular planet.

We have free will. We *are* responsible. We have been provided with eyes and ears to know what is happening. We have been provided with intelligence and intuition to make sense of this information. We have been provided with a sense of responsibility. We have been provided with free will – for good or for ill. And we have been provided with the power to act – the power to save or to destroy, to ruin or to enhance or to husband. We have no excuse for not acting.

However, there are growing causes for hope. Everywhere, in every country, there are people, and groups of people, aware of what is happening and determined to change things for the better. In this book we have tried to encourage other people to follow this path. We have done our best to examine the issues that affect every one of us, to show what damages the world and what does not – what in fact enhances it. And to show, as far as we can with such limited knowledge and experience as we have, what every individual person can do about it.

THE COST OF A HEALTHY EARTH

We live in the most cost-conscious age there has ever been. The science, if it is a science, of economics rules the world. So what would improving the environment actually *cost*?

Well, if the world is to survive we have got to change the science of economics. Dr. E. F. Schumacher subtitled his book *Small is Beautiful*: "Economics as though People Mattered." Perhaps someone should now write a book "Economics as though the World Mattered," because no good is served by saving money if we lose the whole world. We would do better all to become paupers and live on a beautiful and healthy planet than to be millionaires living on a dying one.

But the choice is not actually as dire as that. We do not have to give up all our hard-earned comforts and luxuries and entertainments. We can still lead good lives without polluting and eroding and making deserts. In fact it is generally the case that ecologically sound methods are economically sound too. You don't have to destroy forests to have timber. Properly managed you can have both. And this applies to all economic activities.

So this book is not another doom book – it is a *hopeful* book. Above all it is a book for people, not a book telling what the government ought to do, what society ought to do, what other nations ought to do but what *we* ought to do. Each one of us.

GOOD HOUSEKEEPING IN AN AGE OF WASTE

The agenda for action
Tackling the problem of pollution
Why conservation must begin at home
How to use the power of the purse
Six principles for ecological living

Most inhabitants of the developed Western nations are aware by now that the demands that we are making on our planet are excessive and that our present course is unsustainable. Most of us have an uneasy feeling that all is not well aboard spaceship Earth and that we, as individuals, should really do something about it.

But there are two considerations that many of us feel exonerate us from doing anything. The first is that we think the present state of abundance will last for our lifetime at least. There can be no one who does not realize that our oil supplies will run out one day. But they surely won't run out in *our* time? And if we have children – well, what happens after we

have gone is their worry, isn't it? The second consideration is that in a world which already has an estimated population of four billion souls, surely what one individual does or does not do cannot matter a damn?

Well, the morals that lead us to be swayed by these two considerations are not in question. They are simply deplorable. No system of ethics that has ever been devised can find them anything but squalid. But moral arguments aside, the expediency of being swayed by such considerations is also very much in question.

Such is our human makeup that we can only be truly happy if we are living in harmony with the rest of the world around us. But the sort of life that most people in the developed countries live today revolves around consumption and waste, and this is anything but harmonious. On the contrary, it is highly destructive. If we want to be happier and more contented we need to restrain our demands on the nonrenewable resources of the planet and stop acting in ways that we know perfectly well are tending towards disaster. For this perfectly selfish reason (if no other reason can weigh with us) we should try not to contribute to the process of destroying the only home we have.

A WORLD OUT OF BALANCE

Anyone who has brewed beer knows that yeast is an organism that lives and proliferates by consuming sugar and excreting alcohol, and that in the end it is finally killed, poisoned by its own excrement. When the alcohol content of the beer reaches a certain point most of the yeast will die (although some will sporulate and survive in a dormant form) and the processes that cause fermentation in that tub of beer will cease.

There is a clear parallel between the yeast in a tub of beer and mankind in the tub of the Earth. Modern industry survives by consuming hydro-carbons (oil, coal, and gas) and excreting carbon dioxide. The proportion of carbon dioxide in our atmosphere has risen dramatically since the invention of the steam engine and it is rising at an ever-increasing rate at this very moment. But plants, especially trees, work the other way around. They consume carbon dioxide and excrete oxygen. In this way terrestrial life is sustainable – provided there are enough plants.

But modern man has created two exacerbating factors. The first factor is that he is releasing the vast amount of carbon that was locked up millions of years ago by the plants of the past. Every ton of coal we burn releases carbon dioxide into the air. The second factor is that the forests of the world – by far the most important air purifiers on the planet – are being destroyed at a terrifying rate. Two hundred square miles of forest are being burned or cut down with every day that goes by and are not replanted. Half of all the wood that is felled is burned, thus further adding to the atmosphere's load of carbon dioxide. Quite obviously if we continue on this

course we will go the way of the yeast when the beer becomes too strong.

THE PROBLEM OF POLLUTION

Yeast destroys itself by a single act of poisoning, but the ways by which humans threaten to destroy themselves are abundant and varied, and they do not only threaten our planet in the far future. The most significant of these short-term threats is the way in which we are farming our land. Vast quantities of poisons are dumped on to the soil every year and the "heart," as farmers call it, is going out of the land. We can continue to grow crops on debilitated soil simply by dumping ever-heavier loads of artificial fertilizers on it but this cannot continue forever. Soil is created from rock at a rate of two-thirds inch every four hundred years. On nearly all the arable land of our planet it is being lost at a thousand times that rate. Soil erosion is now accelerating out of control and, as we shall see later in this book, if unrestrained it will become catastrophic.

The substances that are being forced by the chemical industry – with the help of massive advertising – on farmers, foresters and horticulturalists are steadily, and in some places perhaps irreversibly, poisoning our planet. And yet the chemical industry keeps on growing and keeps on turning out its pernicious wares. It is the industry that disposes of the strongest political lobby in every industrial country and it has got completely out of control. Or perhaps it would be truer to say that it is completely *in* control. We all hear of disasters such as Bhopal but we may not all be aware of the never-ending stream of poisonous substances – solid, liquid, and gaseous – that this international tyrant is pouring out upon us.

In addition to this, noxious fumes from power stations, factories, motor cars and even our own homes are causing the phenomenon known as acid rain. In the past, the debate about the exact mechanism by which acid rain damages trees has been used to delay action on countering its effects. But action there must be, and soon, for otherwise no forest – not even in the wildernesses of North America – will be safe in the future. Certainly nobody who has looked at the forests of West

The legacy of consumption
Water and air pollution (*left*) have made huge areas of the industrialized world unfit for forests or agriculture. As well as the visible pollution produced by smoke and waste, there is the potentially more dangerous invisible contamination caused by modern toxic chemicals.

A victim of pollution
Every year millions of seabirds like this guillemot (*below*) are engulfed by oil slicks. Some are rescued and cleaned up, but far more perish unnoticed. Where wildlife and industry meet, the result is often a disaster for animals.

Germany and Eastern Europe can harbor any illusions about this tragedy. If we continue this pollution at the present rate there will be scarcely any trees left to worry about in only a few decades' time.

Acid rain is also affecting our freshwater lakes and rivers. Perhaps the ever-increasing acidity of our waterways has only a marginal interest for most of us: after all freshwater fish do not make up a big part of our diet. But quite apart from the fact that damage to nature will always, eventually, damage ourselves, surely we do not wish to condemn our children to live in a sort of desolate global slum?

WHY WE HAVE TO HELP OURSELVES

Now if the *will* was there, all of these abuses could be stopped without causing any of us to suffer very much. As the following chapters will show, there is no shortage of solutions. For example, modern methods of farming which destroy the soil are mainly there because of shortage of labor. With tens of millions of people unemployed throughout the industrial world, surely labor should not be a problem? Most chemical poisons had not even been invented fifty years ago and yet the world had gone on perfectly well without them. The emissions that cause acid rain could be completely stopped by fitting appropriate filtering devices to the chimneys of plants and into car exhausts. Every problem has its answer.

In theory, the governments of the world could quite easily halt the whole dangerous and dismal progression by passing laws and arranging tax incentives and deterrents. But there is not the slightest chance that any of them will do so voluntarily. There are two sorts of governments in the world: democracies and dictatorships. Most democratic governments are elected every few years. This being so, they will never take any action that is going to lessen their present popularity in order to bring about changes that will only take effect after they leave office. Their only real consideration is re-election.

The generally enlightened government of Ireland provided an example of this when it decided against smoke filters at an enormous new power station planned at a place called Moneypoint. The government argued that the installation of these would put up the price of electricity. Obviously, whichever party was in power, it would never court the unpopularity of increasing the cost of living *now* so that, in fifty years' time, the forests and lakes of England and Scandinavia would not be suffering from pollution. The only democratic governments in the world that have taken any measures at all to clean up pollution have done so only after the fiercest pressure, including direct action from environmental groups, and even then they have often only done so half-heartedly.

Dictatorships have an even worse record. The Russians may have released a few photographs and reports following the Chernobyl disaster, but what of other disasters – the gross soil erosion of the black earth grain belt, or the fearful salinization of the land further east? What of the pollution caused by the heavy industry there, and in Poland and Czechoslovakia? Silence. Those in power in such countries are well aware that the people whom they rule are already disgruntled by constant shortages

and so present production can never risk being exchanged for future environmental benefits. As for pressure from environmental groups – well, that is one thing they do not have to worry about.

We cannot hope that any government, either of the East or the West, is going to take any really effective action to save the future of life on this planet without being forced into it. Almost by definition the people in power have to devote all their energies to consolidating their own position. So we can – and must – try to put pressure on our governments. Where there is a political group with a strong sense of environmental responsibility then we should seriously consider giving them our vote. We should support pressure groups who lobby politicians for change. We should write to local politicians, voicing our dismay and disapproval at things that may be going on locally, nationally and even internationally. But most important, we should ensure that in our own day-to-day lives we make informed choices that in the long term will help rather than harm our planet.

THE POWER OF THE PURSE

Even though as individuals we may not have very much immediate power, at least we all have one piece of muscle that nobody can take away from us. And that is the power of the purse. There is an expression that says with considerable truth "the hand that rocks the cradle rules the world." Perhaps we could reword it thus and still speak with truth: "the hand that rules the purse-strings rocks the world." Every one of us, if we have any money at all, can influence the course of history. If we buy things the production of which, or the disposal of which, causes pollution, then we ourselves are polluters and there is nothing else to be said about it. On the other hand, if we refuse to buy things that are contributing to the destruction of our planet, then we are refusing also to contribute unwittingly to that destruction. And when enough of us are all refusing to contribute in whatever small way, then the destruction itself will stop.

Using the power of the purse needs some straight thinking to counteract the feeling that whatever we do, we cannot do right. It sometimes seems that nearly everything we buy is polluting. Just as an

THE ENVIRONMENT UNDER ATTACK

The global environment can be broken down into three main elements – land, water, and air. Through our daily activities, we manage to pollute and contaminate all three different elements. If it continues, the damage caused by this may become irreversible.

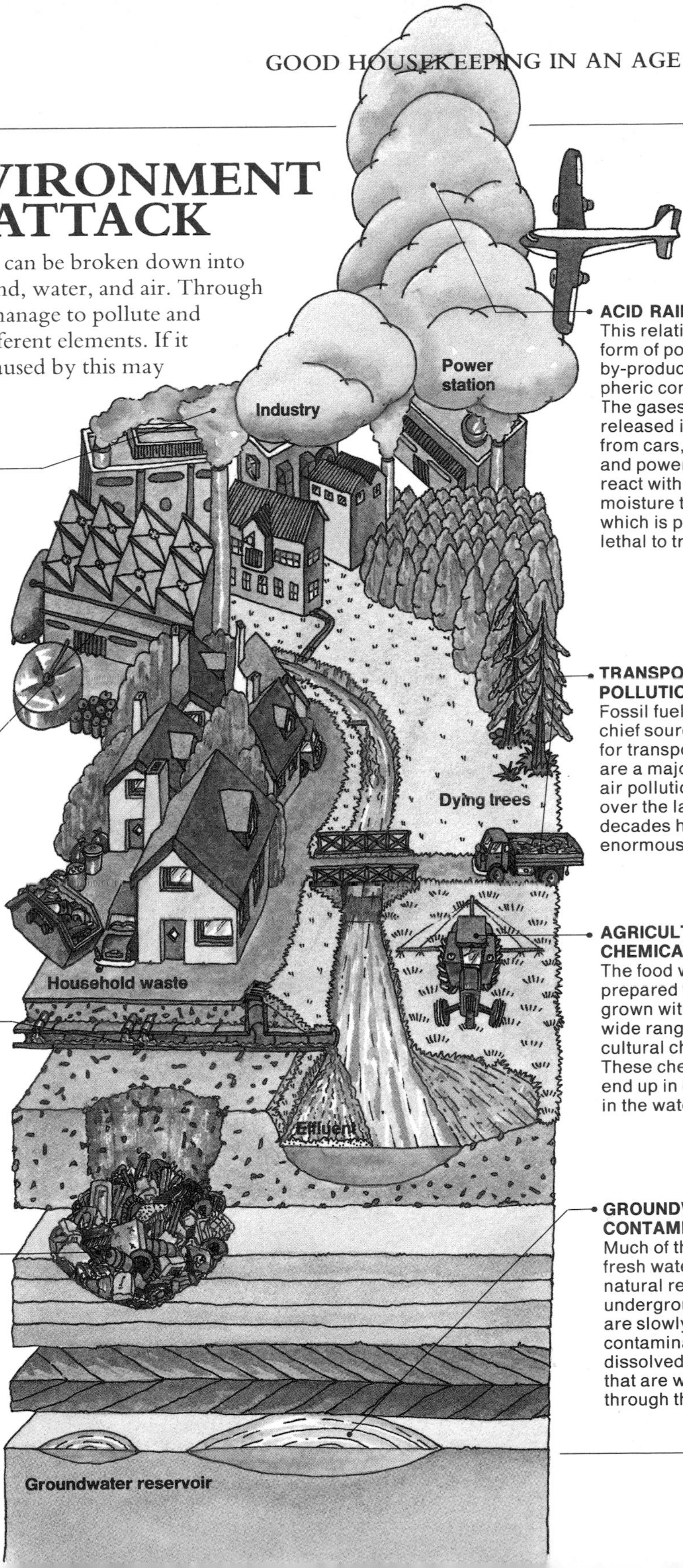

ACID RAIN
This relatively new form of pollution is a by-product of atmospheric contamination. The gases that are released into the air from cars, factories, and power stations react with atmospheric moisture to form rain which is potentially lethal to trees.

POLLUTION FROM ENERGY PRODUCTION
The need for large supplies of energy generates pollution on a huge scale. Energy derived from fossil fuels contaminates the atmosphere, while nuclear power threatens air, water, and land.

INDUSTRIAL AIR POLLUTION
The manufacture of many of the products we use in our homes especially plastics – produces serious pollution, as industrial chemicals are released into the atmosphere.

TRANSPORTATION POLLUTION
Fossil fuels provide the chief sources of energy for transportation. They are a major source of air pollution, one which over the last two decades has grown enormously.

AGRICULTURAL CHEMICALS
The food we eat is prepared from crops grown with the aid of a wide range of agricultural chemicals. These chemicals may end up in our food and in the water supply.

WATER POLLUTION
Water is polluted both by industrial and by domestic users. Like air, water disperses contaminants so that they affect a wide area, and reach places far from their point of origin.

POLLUTING THE SOIL
The burial of household and industrial waste pollutes the ground, and also produces chemical run-off which may reach the groundwater reservoirs that are used for public water supplies.

GROUNDWATER CONTAMINATION
Much of the world's fresh water lies in natural reservoirs deep underground. These are slowly becoming contaminated by dissolved chemicals that are washed through the ground.

Power station

Industry

Dying trees

Household waste

Effluent

Groundwater reservoir

American film actress once said "everything I do is either immoral or fattening," so we may feel that everything we do is helping to ruin the planet. We may feel that it is a hopeless case and not worth bothering with.

The cure for this negative attitude is to become better informed. For it is not true that everything we buy, or do, is harmful to the world. In fact many things that most of us do are good for it. There is no need for despair. We cannot all of us become instantly, miraculously perfect. But we can try.

A FOUR-POINT PLAN

There are four angles from which to look at every action that is open to us. First, is its effect good, bad, or neutral, on our living planet? Second, if its effect is good, how can we encourage and extend it? Third, if its effect is bad, how can we do without it? Fourth, if we think we *cannot* do without it, then what action can we take to reduce or mitigate its damage?

In many cases it is not too unacceptably painful to give up some deleterious action. Do we, for example, have to use half a giant family-size bottle of detergent every time we wash a few dishes? The stuff will all end up in the nearest river and help poison fish, and more than a trace of it will no doubt go down our gullets because we have not managed to rinse it all off the cutlery (although we may think we have). It will do us far more harm than a tiny smear of grease would have done. Of course we don't have to use these quantities. A quarter the amount of the stuff would have done the job just as well. Furthermore there are now perfectly harmless and effective substitutes on the market, and we could perfectly easily use those. By switching to them, the power of the purse will help prevent pollution, with little effort being required.

But what happens when we simply *have* to take some action even when we know it is having harmful effects? Well, we are all desperately worried about maintaining what we call our "standard of living." In developed countries this often means keeping up with the people next door. The time is coming when the most landlubberly of householders will simply have to have a plastic yacht perched on a car trailer in his front drive.

BAD HOUSEKEEPING

Liability for pollution of our planet does not solely rest with the great industrial combines whose factories spew chemical waste into the air, on to the land, into rivers, and into seas. These are all major contributors, but ultimately it is the householder who maintains the demand for their products, and who therefore bears an equal responsibility for the pollution that threatens to ruin our planet.

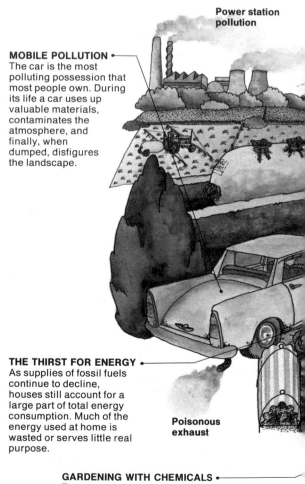

Power station pollution

MOBILE POLLUTION
The car is the most polluting possession that most people own. During its life a car uses up valuable materials, contaminates the atmosphere, and finally, when dumped, disfigures the landscape.

Poisonous exhaust

THE THIRST FOR ENERGY
As supplies of fossil fuels continue to decline, houses still account for a large part of total energy consumption. Much of the energy used at home is wasted or serves little real purpose.

GARDENING WITH CHEMICALS
Too many gardeners rely on chemicals to control what they see as pests. As well as contaminating water and food, these poisons have an injurious effect on harmless garden wildlife. Anything that destroys one form of life will damage another.

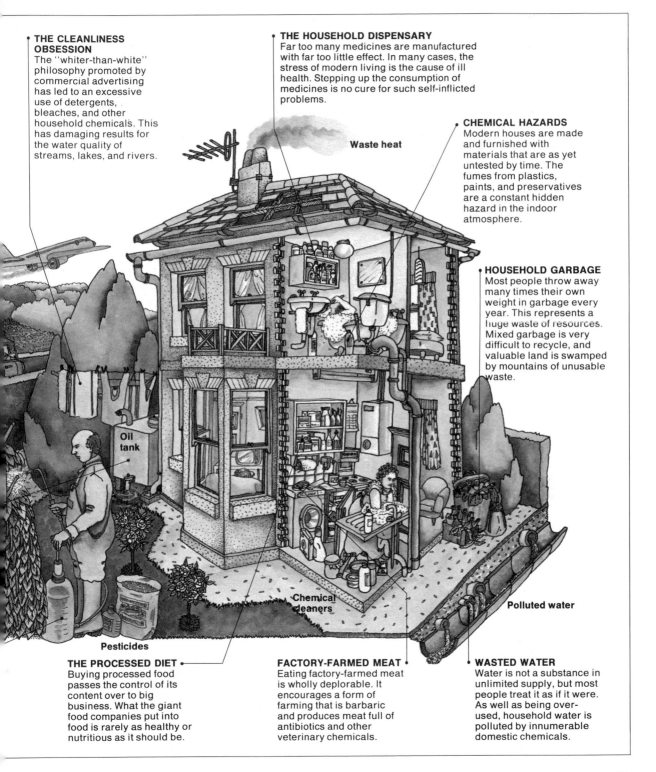

THE CLEANLINESS OBSESSION
The "whiter-than-white" philosophy promoted by commercial advertising has led to an excessive use of detergents, bleaches, and other household chemicals. This has damaging results for the water quality of streams, lakes, and rivers.

THE HOUSEHOLD DISPENSARY
Far too many medicines are manufactured with far too little effect. In many cases, the stress of modern living is the cause of ill health. Stepping up the consumption of medicines is no cure for such self-inflicted problems.

CHEMICAL HAZARDS
Modern houses are made and furnished with materials that are as yet untested by time. The fumes from plastics, paints, and preservatives are a constant hidden hazard in the indoor atmosphere.

Waste heat

HOUSEHOLD GARBAGE
Most people throw away many times their own weight in garbage every year. This represents a huge waste of resources. Mixed garbage is very difficult to recycle, and valuable land is swamped by mountains of unusable waste.

Oil tank

Chemical cleaners

Polluted water

Pesticides

THE PROCESSED DIET
Buying processed food passes the control of its content over to big business. What the giant food companies put into food is rarely as healthy or nutritious as it should be.

FACTORY-FARMED MEAT
Eating factory-farmed meat is wholly deplorable. It encourages a form of farming that is barbaric and produces meat full of antibiotics and other veterinary chemicals.

WASTED WATER
Water is not a substance in unlimited supply, but most people treat it as if it were. As well as being over-used, household water is polluted by innumerable domestic chemicals.

POSITIVE ACTION

A four-point plan for improving the environment

- **Assess the consequences**
  Most everyday activities have some effect on the environment: they either improve it or make it deteriorate. A few leave it unchanged. Deciding which category any action falls into is an essential part of conserving the environment.

- **Encourage positive changes**
  If something enhances the environment, it should be recommended to others wherever possible. Individual action is only successful if enough individuals pursue it.

- **Avoid causing damage**
  Many actions, like throwing away excessive amounts of garbage, wasting water, or using pesticides, fall into the category of avoidable damage to the environment. In nearly all cases, this kind of damage can be prevented without any noticeable change in the quality of life.

- **Cut down what cannot be cut out**
  Some actions, like driving a car, are almost impossible to avoid. In these cases, the best approach is to take steps to reduce the damage.

Maybe instead we should re-examine the whole idea of "standard of living." In very many cases our true standard of living would be raised, not lowered, by owning fewer objects, and of the ones we feel we do need, choosing articles that are simple, local, and made of natural materials by good craftsmen and women. Some Americans, wise in their generation, have invented a way of life which they call "voluntary simplicity." The many adherents of this philosophy all have one thing in common: they are contented.

SIX PRINCIPLES FOR GOOD HOUSEKEEPING

This book is intended to look at every aspect of life in the home, to identify how it affects the world beyond the front door, and to point the way to alternatives that can help us to reduce our impact on the world. No one can hope to produce a complete list of all the problems and all the solutions. However, there are some general principles of good housekeeping, or "ecological" housekeeping if you like, that sum up environmentally conscious living.

ASSUMING RESPONSIBILITY

The first of these principles is the principle of responsibility. All it requires is that everyone should accept that he or she is absolutely responsible for what he or she does or does not do. It is not enough to plead ignorance of the final consequence of our actions. We must make it our business to find out how every product that we use is produced and what effect its production – and its final disposal – will have on our environment. We are responsible, no one else is. There is no "they."

KEEPING THINGS LOCAL

We should try to get what we need locally and we should try to dispose of what we produce locally. This is the second principle – the principle of localism. A mistaken emphasis on the so-called "economy of scale" has broken our local economies asunder. Huge trucks thunder across the Alps carrying apples from Italy to Kent, where they grow some of the best apples in the world, and where orchards of Cox's Orange Pippins are being torn up to open the way for mass-produced Golden Delicious. Even huger trucks blast across the Rocky Mountains carrying lettuces from California to New York State where the people could grow perfectly good lettuces of their own. Early broccoli is grown in Pembrokeshire in Wales. It is carried by trucks right across the country to London where some of it is put into other trucks and sent back again to be sold in shops in Pembrokeshire.

We should try to buy local produce even if it is at first slightly more expensive than stuff from far away. If enough of us insist on it the price will eventually come down and the quality improve. What our locality can produce, it should produce. Freight transportation is heavily polluting, and road transportation in particular uses up vast stores of the Earth's scarce resources, creates endless waste, noise and disturbance, and causes more and more beautiful countryside to be bulldozed flat and put under tarmac. Even a cursory, untrained glance at the nearest motorway will show that far too much freight is being shuffled backwards and forwards across the face of the world. It is completely unnecessary. The power of the purse, if it were widely applied, could end this ridiculous practice.

KEEPING THINGS SIMPLE

Simple things tend to be less polluting and less damaging as well as more truly satisfying than complicated ones. This is the principle of simplicity. There are exceptions to this: a sophisticated enclosed stove may be more economical with fuel than a simple open fire, for example, but in general over-complication is foolish and is a waste of resources. Over-elaborate packaging, such as we find on the shelves of the supermarkets in almost every country in the world, is a conspicuous breach of this principle.

SPREADING THE OPTIONS

Or "don't put all your eggs in one basket." This is the principle of multiplicity, and here follows an example of it not being applied. The energy establishment has been able, with ease and deliberation, to demolish the arguments for every possible source of renewable energy. It points out, quite rightly, that sticking up a few windmills is not going to supply the energy needs of a modern industrial country. Neither is harnessing the power of falling water, nor wave power, nor solar energy collection, nor bio-gas, nor geothermal energy, nor insulating our houses better, nor moderating our demands, nor district heating from the waste heat of power stations or factories, nor heat exchangers, nor incineration. None of these things *alone* can solve our energy needs nor compete with nuclear power or power generated by our dwindling coal and oil reserves. But all these things *together* probably could, and they would do it very cleanly. Unfortunately the establishment is staffed by specialists. Specialists by definition can only consider their own narrow field of work which they would like to see ruling supreme, not sharing the limelight with someone else's invention. Multiplicity is abhorrent to them.

Multiplicity is a key element in reducing damage and pollution wherever there are choices, be they in energy, food, or farming. Having all your eggs in one basket is both dangerous and destructive.

Big isn't always better
The mammoth power stations that produce so much of the air pollution in industrialized countries are a testament to the folly of the big-is-better philosophy. As well as producing energy in massive quantities, they also waste it on a huge scale. The steam that billows out of cooling towers, for example, heats up the atmosphere at huge expense without benefit to anyone. On the other hand, smaller, more varied methods of producing energy, locally sited, are more adaptable and can be much more efficient.

POSITIVE ACTION

Six principles for good housekeeping

The six principles summarized here are guidelines that will help to minimize the impact that your day-to-day life has on the environment. They are *not* a set of rules, but rather a checklist to help you decide what causes pollution and what prevents it.

- **Assume responsibility yourself**
  An understanding of the effects of daily living is essential in preventing damage to the environment. This is the principle of *responsibility* – that everyone should find out about the effects that he or she has on the outside world.

- **Keep things local**
  Energy consumption, pollution, and the use of chemicals are all increased when the supply of a daily commodity, like food, is far removed from the point of consumption. The principle of *localism* is that local products should be given preference to ones from further afield.

- **Keep things simple**
  Increasing complexity brings with it increasing waste, energy use, and pollution. The principle of *simplicity* is that something simple should be preferred to something unnecessarily complex.

- **Avoid specialization**
  Environment-conscious living avoids specialization. The principle of *multiplicity* is that it is safer, less wasteful, and less polluting to maintain variety wherever possible.

- **Avoid violence**
  Physical and chemical violence to the natural world inevitably rebound on us as well. The principle of *nonviolence* requires that all actions that treat land or animals violently should be avoided.

- **Be moderate**
  Excessive consumption creates many of the problems facing the environment. The principle of *moderation* maintains that our consumption should not exceed real requirements.

LIVING NONVIOLENTLY

The principle of nonviolence asserts that we have absolutely no right to use any product or substance if its provision entails unnecessary suffering to other forms of life around us. If we are violent to other creatures we have no right to grumble if violence is visited upon us. This need not preclude our obtaining food to suit our omnivorous appetites, so that if we wish to eat meat, fish, or dairy products then we can. But it does preclude obtaining our food, or any other products such as cosmetics or medicines, with unnecessary cruelty or by inflicting unnecessary damage to the biosphere.

The worm may forgive the plow, as William Blake would have it, but it does not forgive the violence of poisoning the soil in which it lives and from which we all come, and on which ultimately we all depend.

LIVING MODERATELY

This does not mean living life in a dull way, but simply using only what we really need. As a tiny example, the principle of moderation can be applied to the energy used in heating. Do we really need so much? Probably we don't. We are just used to it. We heat our offices like saunas so that everybody has to sit in their shirt-sleeves. Humans have a strong natural tendency towards unnecessary warmth and comfort just as they have one towards too much sugary and fatty food. This immoderate living was once an important survival factor in times of enforced austerity. They are not survival factors any more, now that we have warmth, comfort (as well as sugar and fat) in superabundance. We should exercise self-restraint and thus moderate our demands on our planet.

Last, but perhaps most important, another application of the principle of moderation is in retaining a sense of proportion. We are none of us absolutely perfect and it is no good agonizing to the edge of despair if we drive a polluting gasoline-driven car to a pub or bar, thus transgressing the principles of moderation and simplicity; or because perhaps we may have tried to eat one of those disgusting hamburgers without bothering to enquire where the beef came from, thus contravening the principle of responsibility, the most important principle of them all, and probably the principle of nonviolence into the bargain.

We are only human and naturally very imperfect. Excessive zeal-for-righteousness may turn us off altogether. This does not mean that we should not try, but it does mean that we should not despair if we do not always succeed. Maybe, as we get older, we will get better. Who knows?

REAL OR PROCESSED WATER?

Cutting down on overconsumption
Practical water conservation in the home
Getting our waste back to the land
Household water pollution and how to prevent it
Defusing the nitrate time-bomb
Making water polluters pay

The substance that makes this planet so different from any other that we know of is water. Other planets may have some of it, but the Earth should really be called the Water Planet for water covers most of it. This is why it looks blue and white from space: the white is from water vapor and the blue is from water. And of course all forms of life depend on water.

The part of the world's water that we must consider is the freshwater part, for that is what chiefly concerns us as creatures of the land. At least 97 percent of all the world's water is salt. Like nearly all natural processes, the changing of salt water to fresh, and back to salt again, is cyclical. Water is evaporated from the seas by the heat of the sun and it leaves its salt behind. Winds blow it inland where some of it falls to earth as rain where it makes possible the growth of plants – and us. Then, by one route or another, it returns to the sea where it is mixed with salt again.

Man, like other terrestrial animals, interrupts this cycle. We intercept water somewhere on its passage from the sky to the sea, make use of it – almost always polluting it in some way in doing so – and then send it on its journey to the sea again. As every year goes by, the amount of water that gets diverted from the natural cycle steadily grows.

In most temperate countries there is plenty of it: in fact we may sometimes think too much of the stuff rains down on us from the skies. But our present inordinate demands are placing a strain even on this resource. In Britain, for example, too many Welsh valleys have had to be dammed and flooded – villages, churches, graveyards and all – to meet the apparently insatiable demands of the English cities. Recently $1\frac{1}{2}$ million trees were cut down near the Scottish border to make room for yet another reservoir (one that eventually proved unnecessary, but the damage had been done). Drinking water drowns the land.

And a high price, both environmentally and in money, is paid for this superabundance of piped water. Nearly all of it is expensively purified by passing it over filter beds of clinkers, standing it in settlement tanks, and above all, heavily chlorinating it. The acrid taste of chlorine, which modern city people have now got used to but which their grandparents would never have put up with, is now an all-too-familiar characteristic of the so-called purified water we have to drink.

THE WATER CYCLE
Part 1: *Where our water comes from*

All tap water begins life as rain, trickling into streams, and from there into rivers. Although this water may look clean, there is now hardly such a thing as pure water. From the moment it forms rain until it completes its journey to the sea, water is polluted by man's activities.

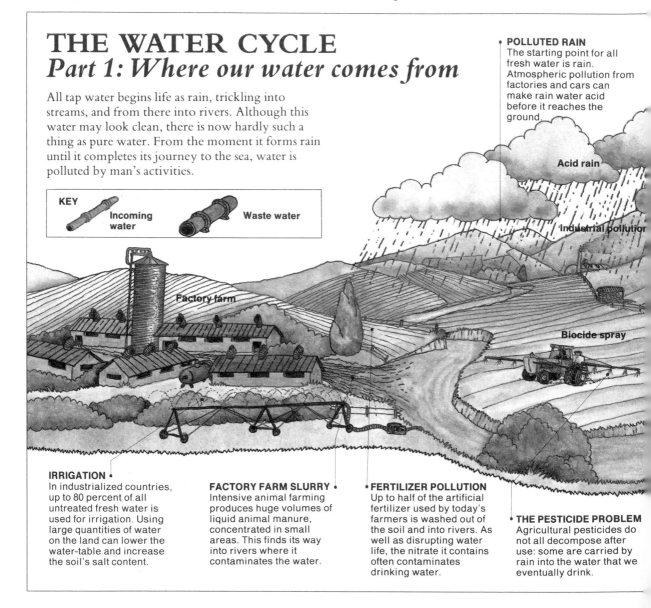

KEY
Incoming water
Waste water

POLLUTED RAIN
The starting point for all fresh water is rain. Atmospheric pollution from factories and cars can make rain water acid before it reaches the ground.

Acid rain

Industrial pollution

Factory farm

Biocide spray

IRRIGATION
In industrialized countries, up to 80 percent of all untreated fresh water is used for irrigation. Using large quantities of water on the land can lower the water-table and increase the soil's salt content.

FACTORY FARM SLURRY
Intensive animal farming produces huge volumes of liquid animal manure, concentrated in small areas. This finds its way into rivers where it contaminates the water.

FERTILIZER POLLUTION
Up to half of the artificial fertilizer used by today's farmers is washed out of the soil and into rivers. As well as disrupting water life, the nitrate it contains often contaminates drinking water.

THE PESTICIDE PROBLEM
Agricultural pesticides do not all decompose after use: some are carried by rain into the water that we eventually drink.

THE COST OF OVERCONSUMPTION

The expense of purifying water, and the sacrifice of good farmland and forests to provide it, is only part of the price we pay for squandering water in the unrestrained way we do.

In all but the countries of very high rainfall, we are, in the matter of water, beginning to live on capital instead of income, and any accountant knows what that means. Many East Anglians can remember when the River Stour, the scene of many of Constable's famous paintings such as *The Haywain* and *Flatford Mill*, began to run so low, because so much of its water was being extracted to wash the cars and flush the toilets of towns like Southend-on-Sea, that there was hardly any of it left. Since that time its flow has been augmented by water pumped (expensively) from underground boreholes.

But what about the underground water supplies or aquifers from which these boreholes extract their

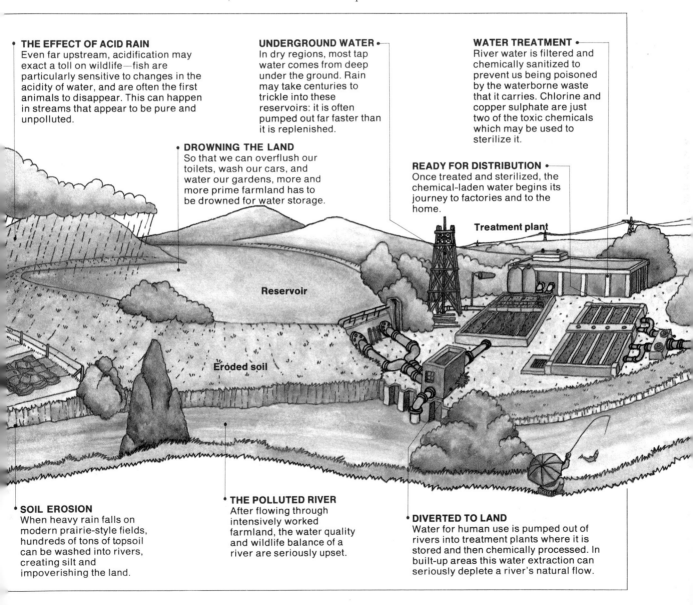

THE EFFECT OF ACID RAIN
Even far upstream, acidification may exact a toll on wildlife—fish are particularly sensitive to changes in the acidity of water, and are often the first animals to disappear. This can happen in streams that appear to be pure and unpolluted.

UNDERGROUND WATER
In dry regions, most tap water comes from deep under the ground. Rain may take centuries to trickle into these reservoirs: it is often pumped out far faster than it is replenished.

WATER TREATMENT
River water is filtered and chemically sanitized to prevent us being poisoned by the waterborne waste that it carries. Chlorine and copper sulphate are just two of the toxic chemicals which may be used to sterilize it.

DROWNING THE LAND
So that we can overflush our toilets, wash our cars, and water our gardens, more and more prime farmland has to be drowned for water storage.

READY FOR DISTRIBUTION
Once treated and sterilized, the chemical-laden water begins its journey to factories and to the home.

Treatment plant

Reservoir

Eroded soil

SOIL EROSION
When heavy rain falls on modern prairie-style fields, hundreds of tons of topsoil can be washed into rivers, creating silt and impoverishing the land.

THE POLLUTED RIVER
After flowing through intensively worked farmland, the water quality and wildlife balance of a river are seriously upset.

DIVERTED TO LAND
Water for human use is pumped out of rivers into treatment plants where it is stored and then chemically processed. In built-up areas this water extraction can seriously deplete a river's natural flow.

water? These are not inexhaustible either. Aquifers all over the world are shrinking through over-consumption of their water. The famous Ogallala Aquifer, which underlies a vast area of the drier parts of the western Great Plains of the United States, now has a life expectancy of about forty years. Its ancient "fossil" water was laid down in Pleistocene times. When it is gone, it will take thousands of years for it to be replenished. As for rivers – the mighty Colorado no longer runs out to the sea. All its water is used up before it has a chance to get there.

As the water-table sinks wells dry up and wild-life suffers. Florida's rapidly growing population is having precisely this effect on the state's magnificent wetlands.

In some cases, the man-made water supply system carries far more water than its natural counterpart. In California, for example, an aqueduct carries 4.2 million acre-feet of water every year over

THE WATER CYCLE
Part 2: How water is used and polluted

Once water has been treated, it is distributed for use. The amount we consume through drinking and eating is completely eclipsed by that used in our homes and factories – and almost every gallon of it is polluted before being thrown away.

WATER'S HIDDEN USES
Factories use huge quantities of water to make consumer products. Up to 120,000 gallons of water can be used in washing, cooling and processing the components that make up a car.

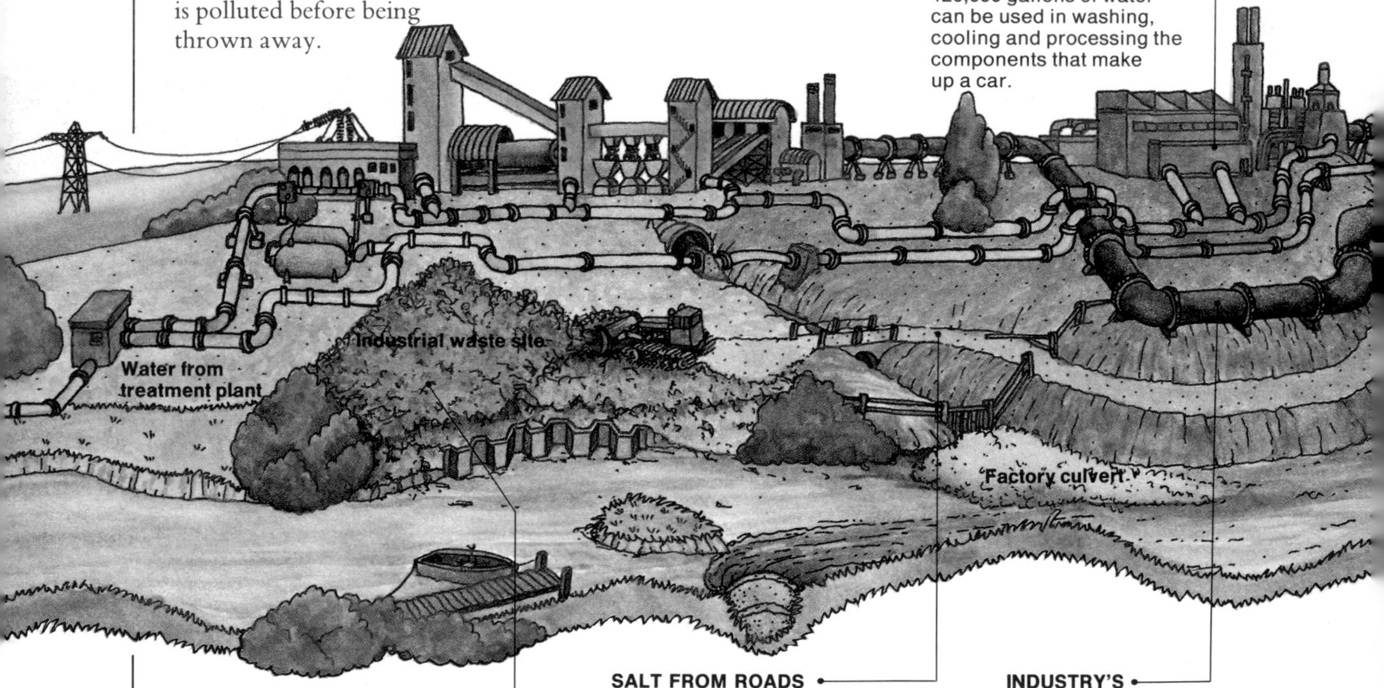

Industrial waste site

Water from treatment plant

Factory culvert

UNDERGROUND CONTAMINATION
Buried industrial metals and solvents can seep through the ground to reach streams and rivers. No one yet knows what long-term effects these will have.

SALT FROM ROADS
Every winter millions of tons of salt are spread on roads where ice is a problem for traffic. Much of this salt is washed into streams and rivers where it harms both animals and plants.

INDUSTRY'S LIQUID WASTE
Most of the water used and contaminated by inland industries is discharged directly into the sewage system, loading it with chemicals which are impossible to remove.

hundreds of miles to supply the cities of Los Angeles and San Diego, and the irrigated farmland of San Joaquin Valley. Two huge pumps lift much of this water over 3,000 feet to enable it to flow south rather than north, consuming enough electricity to power a sizable town in the process. A lot of this water is simply wasted. Ordinary public water supplies may be much less grandiose than California's, but they still account for enormous volumes of water.

With industrialized countries sucking out so much water from urban rivers, the water flow is maintained only by the effluent that is discharged into them. One effect of this is that in highly populated areas effluent is reprocessed into drinking water again and again. It is commonly said that some of the water that runs out of the Thames may have been through six people's bodies before it does so. The same could be said of many other urban rivers: their water is often swallowed many times before it eventually reaches the sea.

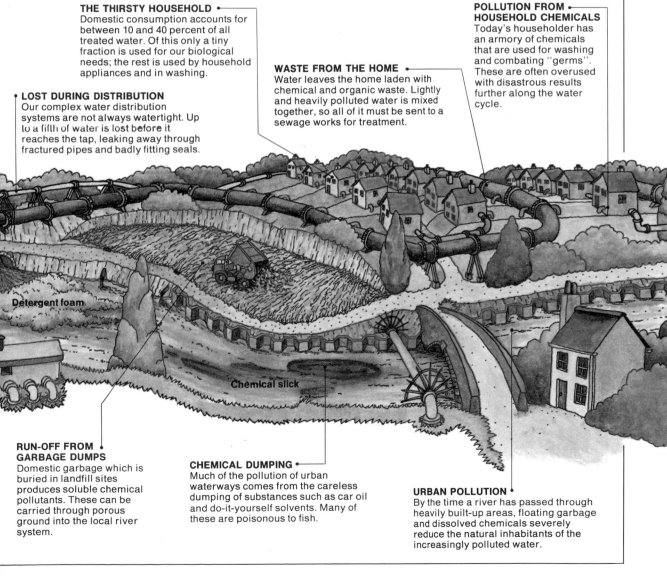

THE THIRSTY HOUSEHOLD
Domestic consumption accounts for between 10 and 40 percent of all treated water. Of this only a tiny fraction is used for our biological needs; the rest is used by household appliances and in washing.

WASTE FROM THE HOME
Water leaves the home laden with chemical and organic waste. Lightly and heavily polluted water is mixed together, so all of it must be sent to a sewage works for treatment.

POLLUTION FROM HOUSEHOLD CHEMICALS
Today's householder has an armory of chemicals that are used for washing and combating "germs". These are often overused with disastrous results further along the water cycle.

LOST DURING DISTRIBUTION
Our complex water distribution systems are not always watertight. Up to a fifth of water is lost before it reaches the tap, leaking away through fractured pipes and badly fitting seals.

Detergent foam

Chemical slick

RUN-OFF FROM GARBAGE DUMPS
Domestic garbage which is buried in landfill sites produces soluble chemical pollutants. These can be carried through porous ground into the local river system.

CHEMICAL DUMPING
Much of the pollution of urban waterways comes from the careless dumping of substances such as car oil and do-it-yourself solvents. Many of these are poisonous to fish.

URBAN POLLUTION
By the time a river has passed through heavily built-up areas, floating garbage and dissolved chemicals severely reduce the natural inhabitants of the increasingly polluted water.

USING WATER MORE WISELY

Human nature being what it is, the most success-ful way of rationalizing our use of water will be one that promises to save us money as well. That is just what could happen if our water supplies were designed more sensibly.

The problem to be tackled is this. Every day we splash, pour and flush our way through between about 40 and 130 gallons of water at home, while even more is used on our behalf in factories. A quarter of the piped water in England is lost through simple leakage out of the ancient and corroding mains. A vast amount is lost through leaky taps. Much is sacrificed to the worship of the great God Car, and more on keeping our lawn greener than that of the Jones' next door.

Of all this water, only a trifling amount, say 30 gallons, actually needs to be drinkable. All the rest, the far larger amount, doesn't need to reach this standard. We certainly need pure water to drink, to

THE WATER CYCLE
Part 3: What happens to waste water

After using and polluting water, we throw it away. It is easy to pull a plug or flush a toilet; getting rid of the waste this produces is a lot more difficult. As waste water approaches its final destination – the sea – an alarming number of the potentially dangerous substances it contains find their way back to land.

CLEANING UP WATERBORNE WASTE
Organic matter in sewage is biologically digested at the sewage farm. This process, which relies on bacteria, can be wrecked by antibiotics, bleach, and other common household chemicals.

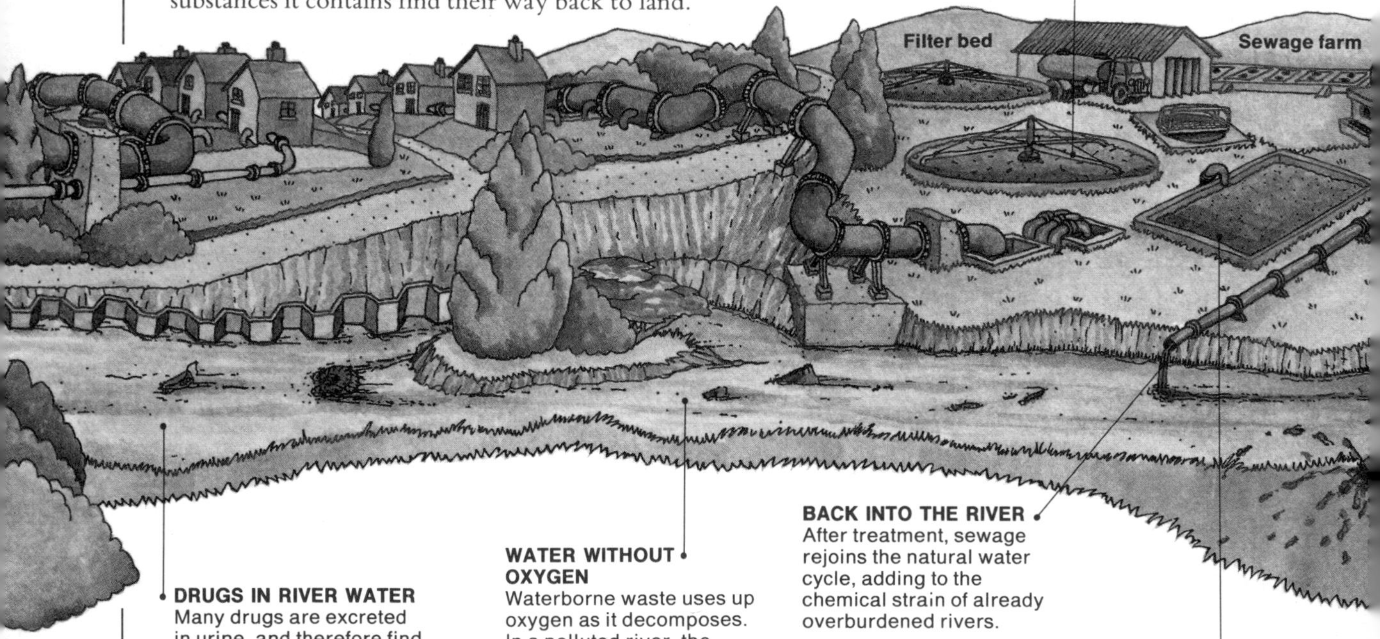

Filter bed

Sewage farm

DRUGS IN RIVER WATER
Many drugs are excreted in urine, and therefore find their way into rivers. Hormones used in contraception are now reaching detectable levels in some urban rivers. Reduced fertility may result if they are passed on into drinking water.

WATER WITHOUT OXYGEN
Waterborne waste uses up oxygen as it decomposes. In a polluted river, the "oxygen debt" may reach a level at which fish cannot survive. In some urban rivers, oxygen is bubbled into the water to keep fish alive.

BACK INTO THE RIVER
After treatment, sewage rejoins the natural water cycle, adding to the chemical strain of already overburdened rivers.

SEWAGE PONDS
In places where sewage sludge is not used on the land or dumped at sea, it is poured into deep "ponds" from where it seeps into the surrounding land.

cook with, and to wash in, but we most certainly don't need pure water to flush the toilet, wash the car, water the lawn or make car tires.

But that is just what happens at present. Nearly everywhere, highly expensive processed water is consumed in uses where its biological cleanliness is quite irrelevant. In some cases, like watering the garden, its content of disinfectant might even be a disadvantage (most plants prefer real water). What is needed is a *dual* water system – two sets of pipes

and two sets of taps in each home and factory.

Obviously, no single householder could embark on this alone, but we can all make it plain that this is what we want. The dual system might sound like a massive investment in plumbing, but it is actually an investment that would pay for itself very quickly. Every country that installed water mains a hundred or more years ago has got to face the reality that, being made of cast iron, entire systems are rusting to pieces and have got to be re-

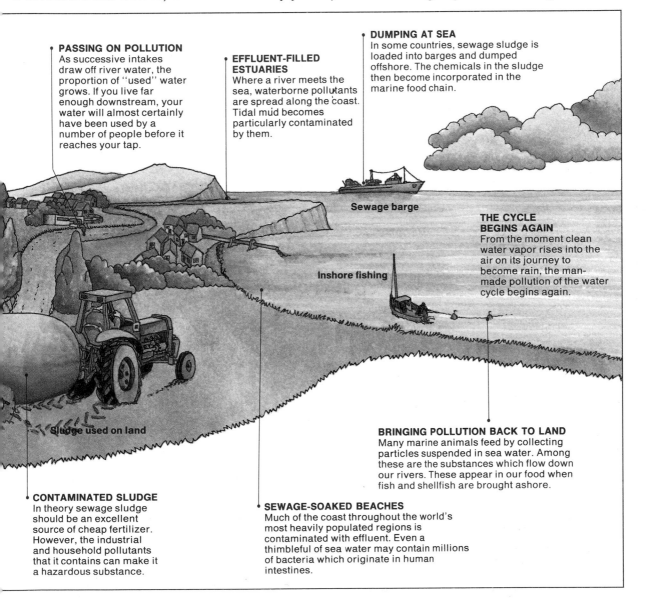

PASSING ON POLLUTION
As successive intakes draw off river water, the proportion of "used" water grows. If you live far enough downstream, your water will almost certainly have been used by a number of people before it reaches your tap.

EFFLUENT-FILLED ESTUARIES
Where a river meets the sea, waterborne pollutants are spread along the coast. Tidal mud becomes particularly contaminated by them.

DUMPING AT SEA
In some countries, sewage sludge is loaded into barges and dumped offshore. The chemicals in the sludge then become incorporated in the marine food chain.

Sewage barge

THE CYCLE BEGINS AGAIN
From the moment clean water vapor rises into the air on its journey to become rain, the man-made pollution of the water cycle begins again.

Inshore fishing

Sludge used on land

BRINGING POLLUTION BACK TO LAND
Many marine animals feed by collecting particles suspended in sea water. Among these are the substances which flow down our rivers. These appear in our food when fish and shellfish are brought ashore.

CONTAMINATED SLUDGE
In theory sewage sludge should be an excellent source of cheap fertilizer. However, the industrial and household pollutants that it contains can make it a hazardous substance.

SEWAGE-SOAKED BEACHES
Much of the coast throughout the world's most heavily populated regions is contaminated with effluent. Even a thimbleful of sea water may contain millions of bacteria which originate in human intestines.

THE WORLD'S THIRSTIEST COUNTRIES

This chart shows how much water is removed from the natural water cycle every day by some of the world's thirstiest countries. It also shows this amount as an equivalent per head of the population. The figures shown are for all uses — industrial, agricultural as well as domestic.

	Daily total (millions of gallons)	Equivalent per person (gallons)
United States	380,160	1,668
Canada	26,400	1,090
Australia	13,200	876
Netherlands	10,296	721
Italy	39,600	710
Spain	26,400	700
Japan	76,560	668
Belgium	6,600	663
Finland	2,904	560
Germany	30,360	494
France	20,592	362
Norway	1,320	352
Sweden	2,904	346
New Zealand	792	277
Great Britain	9,504	185
Denmark	792	172
Switzerland	528	77

placed. Why not grab the bull by the horns, make a handful fewer ballistic missiles for a year or two, and use the money to replace the whole system with two systems: a drinkable supply and a non-drinkable one?

Surely it would be better to have a separate, small-bore system to provide the few gallons of drinking water needed per day to each household and then another system of larger pipes to provide all the other? And both supplies would have to be metered and paid for. The metering is important because it provides the incentive for everyone to cut down their consumption — a sound mixture of self-interest and conservation.

If a dual supply were provided, all of the drinking water (a tiny proportion of the total) would come from formations deep under the ground, which provide the purest untreated water, or else from thoroughly filtered surface water, without chlorination. It would taste nice and be quite safe to drink. The much greater quantity needed for industrial use, washing cars, flushing lavatories and watering the garden would not need elaborate and expensive treatment.

WHAT'S WRONG WITH BOTTLED WATER

The water that most city people have to drink nowadays is highly processed and, quite frankly, disgusting. It is not surprising that more people are turning to bottled water. Much of this comes from

WATER'S HIDDEN USES

A large amount of everybody's daily water consumption is accounted for by the manufacture of products destined for the home. This chart shows how much water a year's supply of some common household items takes up. These figures assume a moderate level of consumption: if you are addicted to canned drinks or read several newspapers, your "hidden" water consumption will soar.

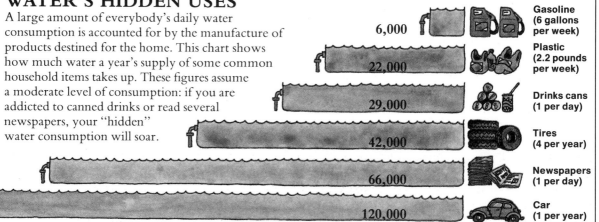

Water consumption (gallons)	Product
6,000	Gasoline (6 gallons per week)
22,000	Plastic (2.2 pounds per week)
29,000	Drinks cans (1 per day)
42,000	Tires (4 per year)
66,000	Newspapers (1 per day)
120,000	Car (1 per year)

underground or glacial sources that are exceptionally pure. To anyone accustomed to chlorinated water it tastes delicious, and, for the people who bottle it, it is a very lucrative product.

The trouble with relying on bottled water as an answer to the poor quality of tap water is that it encourages a wasteful and polluting trade. The waste in moving just water around the world in bottles is appalling. Besides the cost in fuel for transport, there is the cost in packaging. A ton of glass bottles takes the energy equivalent of a ton of coal to make, while plastic bottles are notorious pollutants.

Certainly we could drink bottled water as a luxury, but to treat such a basic substance as a necessity creates a traffic which is surely unacceptable. Investing in domestic water filtering devices and fitting them to taps or water jugs is an effective but costly means of solving the problem. Improved water supply systems are a far better way of spending our money.

PRACTICAL WATER CONSERVATION IN THE HOME

In 1977 during a severe drought in California people were hanging signs up in their lavatories which proclaimed the following sensible (if inelegant) injunction:

"If it's brown wash it down. If it's yellow let it mellow."

They had good reason for this. A quarter of all the domestic water in most countries goes straight down the toilet. Every time somebody flushes the toilet about 5 gallons of water are instantly changed from being pure to being polluted.

If installing a dual water system in your house is impossible at present, there are plenty of practical ways by which you can reduce your consumption of processed water without it making a jot of difference to your standard of living.

The toilet is a good place to start. A simple double-action device installed in the flushing system halves the amount that is flushed away. The conventional toilet in the United States uses 5 gallons of water every time it is used, the common low flush toilet $3\frac{1}{2}$ gallons, the washdown $1\frac{1}{2}$ gallons and the air-assisted $\frac{1}{2}$ gallon, so that there is a potential saving of 90 percent in this one measure.

The large difference in water consumption of a bath and shower is probably well known – one average bath contains enough water to keep a shower going for a quarter of an hour. But household machinery can also be very thirsty. Washing machines and dishwashers are particularly carefree with their use of water, and often this can be reduced at a stroke by using the machines only when they are full, or by using a half-load program wherever possible. This also reduces water pollution because less detergent is used overall.

One old-fashioned water saving device which should be brought back into every home is the water butt. It is true that our rainwater may not be as chemically clean as it was in the past, thanks to atmospheric acidity (see p. 144), but don't let this put you off collecting it. Although it may not be clean enough to drink, it will be perfectly good for cleaning your car or watering the garden. The collective roof area in industrial countries is enormous. In most major cities about one-tenth of the ground is built on: if all these roofs were connected to tanks rather than drains, huge reserves of water would be available.

Saving a few gallons here and there in your home may seem a waste of time, particularly if you live somewhere where pouring rain is more of a problem than drought. But remember, rain water and treated water are two different substances: the liquid that comes out of your tap is prepared using unpleasant and even harmful chlorine. The less treated water we use, the less of this potentially dangerous chemical we need make, and the less we need to have around.

WHAT HAPPENS TO WATERBORNE WASTE

Now what happens to the water when we have finished spoiling it? A quarter of it, as we have discovered, goes down the toilet. The rest goes down various drains and, like the other, ends up in the sewage system. Then, apart from a rising proportion which leaks out of crumbling nineteenth-century brick sewers, it eventually gets to some sort of sewage treatment plant. Here the solids are allowed to settle out and the liquids – after bacterial action – are pumped out into the nearest

WATER IN THE HOME

This chart shows how water is used once it arrives in the home and what potential there is for reducing the quantities used. The bathroom accounts for the largest amount and for this reason it is the best place to begin water conservation. The toilet accounts for slightly less than the bathroom, while about half of the water supplied is used in washing clothes, in the kitchen, or outside in the garden.

Washing and bathing	Toilet	Laundry	Dishwashing	Drinking and cooking	Outside use
27%	24%	17%	14%	10%	8%
Potential for reduction High: can be reduced by use of shower.	**Potential for reduction** Very high: can be reduced by water-saving design.	**Potential for reduction** Moderate: can be reduced by keeping to full loads.	**Potential for reduction** Moderate: can be reduced by keeping to full loads.	**Potential for reduction** Low: these essential uses cannot be reduced.	**Potential for reduction** High: most outside uses are non-essential.

river, probably to be extracted lower down and used again. The solids, still fairly obnoxious, after a drying process, are then either dumped in a landfill site, or, more and more, taken out to sea. Incineration has been tried but is expensive: sea-dumping is the preferred method wherever practicable. To take England and Wales as an example, very roughly one-fifth of the sewage sludge produced is dumped at sea, two-fifths dumped in landfill sites and two-fifths applied as fertilizer to agricultural land.

Now here we come to a question of enormous importance to every one of us, or at least to our dependants. This planet's buried store of natural plant nutrients – nitrates, phosphates and potash – is strictly limited. As we shall see in the next chapter, guano (the part of this store that was created by fish-eating birds over tens of thousands of years) was used up by Western man in less than a hundred years and simply dumped, in sewage, back into the sea, from which it is irrecoverable. Mankind has practically destroyed the birds' fish supplies, so that is the end of that supply.

But the nitrates, phosphates and potash that these deposits contained could be working for us *now*. They could still be growing crops! For, with good husbandry, plant nutrients do not always get used up. They can be recycled again and again, more or less forever. But they were put on the land, taken up by crops, these were eaten, and all the nutri-

ents not absorbed by humans went down the toilet.

The bonanza of the last century has been blown. However, since that time large deposits of rock phosphate have been discovered (very often the fossilized remains of living creatures in the past), together with large deposits of potash (left over from the drying-up of seas geological ages ago) and, above all, huge deposits of oil and natural gas.

Now nearly all of this – except the oil and natural gas – we are currently dumping into the

POSITIVE ACTION
Simple ways to save water

- **Choose water-efficient appliances**
  There are large differences in the amount of water used by different brands of washing machines and dishwashers. Choosing appliances that are sparing with water will reduce your consumption.

- **Cut down on car-washing water**
  A large proportion of the domestic water that is used outside the home gets consumed in washing cars. Although a car that is regularly washed will generally last longer, nothing is to be gained by overdoing it: obsessive car-washing just wastes water.

- **Recycle your kitchen water**
  Dirty dish water is generally harmless to plants. Using it to water your vegetable garden will save tap water in summer.

sea. We are burning most of the oil and natural gas of course, but much of it goes into making artificial fertilizers. And all of this bounty, this wealth, gets dumped into the sewers, into the rivers, and ultimately into the sea. And neither we nor our descendants will ever get it back again.

And so there is a strong demand, from people who realize what is happening and – can we say without too much arrogance – who can possibly see further than their noses, that this process should be halted. The metabolism of our species is at present *linear*. Phosphates, to take one example, come out of North African phosphate mines, are shipped to European or North American farms, are used to grow one crop, and then go down the sewer out into the depths of the sea where they will remain.

But it should be *cyclical*. Before the invention of the water closet it was. The phosphates that occur naturally in all soils were slowly released, carefully composted and taken back to the land whence they came so the cycle could start again.

WHY WATERBORNE WASTE IS THROWN AWAY

We people who object to sewage being dumped into the sea come up against what is apparently an immovable object. And that is the incontrovertible testimony of the scientists – honest and dedicated people no doubt – who actually have the job of removing our sewage. And here are their arguments.

First, that there are so many poisons now in sewage, heavy metals like lead and cadmium – the latter mostly from rust-proofing of sheet steel and making orange and yellow plastics, and the former mostly from lead added to gasoline – that most sewage would be dangerous if put on the land.

Second, and this is by far the most potent argument, that the sludge left after the necessary treatment of sewage contains so little nitrate, phosphate or potash, that whether or not it returns to the land hardly makes any difference at all.

Now, to take England and Wales as one example, if the million or so tons of dried sewage sludge which is produced every year were returned to the land, it would only represent a mere 4.5 percent of our present requirements of nitrate and phosphate and an even merer 1 percent of our potash. The small amount of nutrients returned would simply not justify the danger presented by the pathogens (disease-causing organisms) that might survive the treatment process, or the cost of removing the manifold poisons that now go into our sewage.

Now very little of the nitrate from artificial

The sewage business
In a large urban sewage treatment works a great quantity of time, money, and space is made over to dealing with a substance simply so that it can be thrown away. In industrialized countries sewage is seen as something to be gotten rid of at all costs, regardless of the fact that it contains useful natural fertilizers which should go back to the land – not into the sea.

fertilizers gets into sewage. A certain proportion of it is taken up by the crop but much of it is simply washed into the soil to pollute the water supplies. But unlike nitrate, the phosphate and potash are not quickly washed into the soil. They are either taken up by the plants or retained by chemical reaction with other soil elements to be used again. However, of the quantity that is put on the land every year, only about 5 percent turns up in sewage. The rest apparently vanishes.

So why is it that vast amounts of phosphate and potash do not, as one would expect, appear in sewage? Where have they gone? The answer is that they are soluble, and therefore do not get left in the sewage sludge at all. They are washed out in the huge amounts of water that run through our sewage plants into rivers and into the sea. The figures that the water engineers give us are for sewage sludge only and therefore are the wrong figures, because the amount of nutrients left in sludge is minute in relation to the amount that is dissolved and lost.

Of course the sewage engineers have been asked to do the wrong thing. They have been told to "get rid of waste". Instead of this their brief should have been entirely different. It should have been to "recover for the use of agriculture the extremely valuable materials that occur in human sewage". To rebuild the cycle in other words, the cycle of soil–plant–animal–man–soil again. Instead we now have a linear system: phosphate and potash mines–soil–plant–animal–man and down the lavatory to the sea. If we can have mines–soil–plant–animal–soil again, eventually the mines would become redundant.

GETTING OUR WASTE BACK TO THE LAND

Now to achieve this a completely new system of sewage disposal will have to be introduced. It will have to be a dual system, where human sewage is kept separate from the start from all other wastes.

Fortunately, the sewers we have now, in all the older industrial countries, are near the point of collapse. They will have to be renewed anyway. And when they are renewed we must insist that they are renewed in a completely different way.

Since 1959 a method of sewage disposal has been used in Sweden, and has spread to many parts of the world. It has proved to be completely successful. Air is used instead of water to transport sewage, and the latter is separated into what the Swedes call "black" water and "grey" water. The grey water consists of all liquid wastes from the home that are *not* sewage. It is conveyed in separate pipes to the sewage treatment plant, rendered safe, and discharged into a river. The black water is actual sewage, with the nutrients in it. Compared with the grey water, this is only a comparatively tiny amount of material. And, after treatment, simple and easy, it is put back on the land. But as this airborne system is expensive, a modified water-borne one, using less water, would do as well.

Now undoubtedly the best treatment this sewage could receive is to be put through a methane digester. This method is so straightforward that there is very little to say about it. Simply it is this. Organic matter is put into an airtight tank in which anaerobic bacteria (bacteria which live without air) get to work on it. As they digest they produce methane gas. This is drawn off and put to any use which, for example, natural gas is put. It is, in fact, a natural gas. The spent sludge, with the gas removed from it, is left as a completely safe, sterile and highly valuable fertilizer. By far the best thing to do with it then is to compost it – that is, to mix it with low-grade organic waste like straw, municipal waste or anything organic. The nitrogen in the sludge (no nitrogen is lost in the methane) will feed the putrefactive bacteria to enable them to break down the organic waste. At present astronomical tonnages of straw are being burned in most Western countries: this could be used. The resulting manure, or compost, is pleasant-smelling, completely safe, and can be returned to the soil as a marvelous manure.

In China there is a methane plant in nearly every village, and they work perfectly. The village is provided with free lighting and heating, the land with a fine fertilizer, and what our engineers would call "waste" is removed without any unpleasantness to anybody. What we have done in Western countries is to make three problems out of the water cycle: the problems of how to fertilize the soil, how to prevent pollution of the rivers and seas,

THE DUAL WATER SYSTEM

This diagram shows how a house's plumbing could be arranged to make the best use of both water supply and waste. There are two supply pipes. One provides pure drinking water to the kitchen; the other provides treated water to all other taps and appliances. Waste water leaves the house through two separate pipes. One carries sewage which is returned to the land after composting, while the other contains used water which is returned to the water cycle.

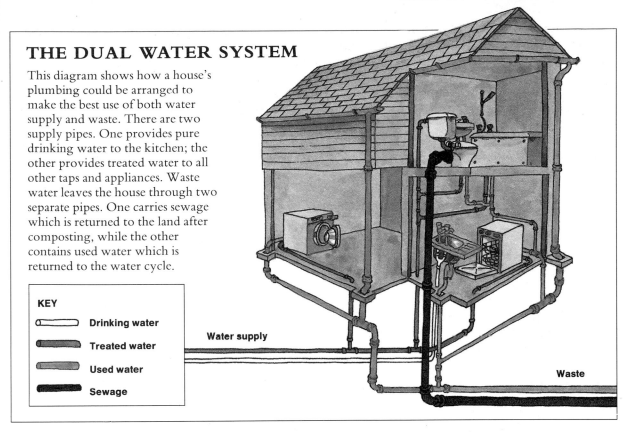

KEY

- Drinking water
- Treated water
- Used water
- Sewage

Water supply

Waste

and how to get rid of sewage. By mining out the stored nutrients of the Earth and, after using them *once*, dumping them into the sea where they are forever out of reach, we are robbing our posterity.

HOUSEHOLD WATER POLLUTION AND HOW TO PREVENT IT

One of the chief obstacles that stands in the way of wholesale effluent recycling is the fact that it is often poisonous. Most people are probably aware of the problems of industrial effluent, but fewer realize that it is not only uncaring big business that is responsible for water pollution: the ordinary householder can also poison or pollute the water cycle. Action on this front therefore begins with putting our own houses in order.

It is ironic that, having put so much effort into taking pollutants out of our tap water, we start putting them back in as soon as the water reaches our homes. Of all the water used in the home only a tiny fraction, perhaps 5 percent, escapes some form of chemical adulteration caused by detergents, bleaches and scouring powders.

Until the 1930s the only man-made chemical that went out with the dishwater was soap, a fairly innocuous substance. Even in moderate concentrations it is harmless to life in lakes and rivers. Soap has since been superseded by a range of powerful synthetic substances, commonly found in dishwashing liquid, laundry detergents, bubble baths, shampoo, and cosmetics.

A squirt of liquid detergent in your dishwater adds one of the most difficult pollution problems currently affecting the water cycle – eutrophication, or the overabundance of chemical food in water. Detergents may not sound much like food, but they often contain phosphates, which for plants are important growth materials. Algae – minute waterplants – run riot in water with high levels of phosphates. When phosphate-rich dishwater runs down the drain and into the waterways, the algae pounce on this abundant supply of nourishment.

The result is eutrophication – previously clear water begins to resemble something much more like pea soup: thick and green.

This may sound harmless, but for many water organisms it has terrible side effects. Water life needs dissolved oxygen. In eutrophic water, plants take up so much of the oxygen that there is little left over for other water life. Fish and other animals begin to suffocate. The plants continue to multiply and rot down, and eventually the water becomes clogged and lifeless.

There is absolutely no need for this to happen because detergents without phosphates exist and are effective. Proof that alternative products are workable comes from Switzerland, where phosphate-containing detergents have been banned altogether. Alarmed by the eutrophication of their lakes, the Swiss decided to do something about it. Other countries should follow suit.

POLLUTION FROM SOLVENTS

A second area of household water pollution concerns our use of oil-based products and solvents. One thing that is guaranteed to harm freshwater life is a film of impermeable oil floating on the surface, cutting the water off from the atmosphere. Three-quarters of all the oil contamination at sea is estimated to come from land, and although much of this is from industry, the householder who tips a half-used can of old gloss paint down the outside drain is equally guilty of appalling thoughtlessness. Paint, turpentine, and all other oils and solvents should *never* be poured away like this. If you really have to dispose of them, the safest place is a very deep hole – below the topsoil/subsoil boundary – and well away from any trees.

Also in this catalog of contaminants come garden poisons (see chapter 9). Anyone who feels that they must use these should take special precautions to see that they do not reach flowing water – this means keeping them well away from both streams and drains.

THE NITRATE TIME BOMB

We have lived for a long time with the good old-fashioned poisons in our drinking water such as arsenic, lead, cadmium, and the rest of it. Water

WATER POLLUTION BEGINS AT HOME

Water pollution is not something created simply by factories run by uncaring big business. All of us are to some extent responsible for the other side of water pollution – that created by the carefree use of detergents, bleaches, scouring powders, and a host of other household products.

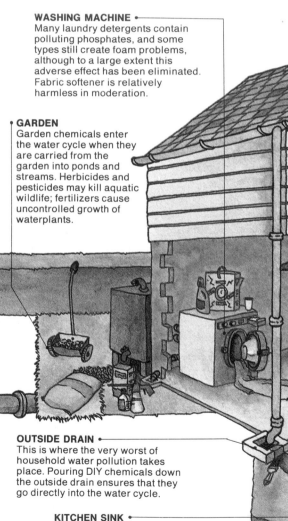

WASHING MACHINE
Many laundry detergents contain polluting phosphates, and some types still create foam problems, although to a large extent this adverse effect has been eliminated. Fabric softener is relatively harmless in moderation.

GARDEN
Garden chemicals enter the water cycle when they are carried from the garden into ponds and streams. Herbicides and pesticides may kill aquatic wildlife; fertilizers cause uncontrolled growth of waterplants.

OUTSIDE DRAIN
This is where the very worst of household water pollution takes place. Pouring DIY chemicals down the outside drain ensures that they go directly into the water cycle.

KITCHEN SINK
Liquid dishwashing detergents are often a serious source of phosphate pollution. Scouring powder contains bleach which is poisonous to the bacteria which decompose sewage.

TOILET
The chemicals most often flushed down the toilet are bleach and naphthalene-based water "fresheners". Both of these can poison the sewage-treatment process.

BATH AND SHOWER
The soap used in washing is generally harmless to the water cycle (it consists mainly of vegetable oils) but shampoo can contain large amounts of phosphates, which cause water eutrophication.

POSITIVE ACTION
How to reduce water pollution

● **Liquid detergents**
The main problem with these is their phosphate content. Use phosphate-free or low-phosphate dishwashing liquid if you can obtain it. If not, try reducing the quantity you use.

● **Laundry detergents**
No one with a washing machine can avoid using these. However, try experimenting with reduced quantities – remember manufacturers have an interest in making you use more than is actually necessary.

● **Bleach and scouring powders**
Again, the problem here is one of quantity. Very dilute bleach left to work for longer is just as effective as a strong solution poured away after a few minutes.

● **Water fresheners**
Don't use them. They do not "freshen" water at all – rather they pollute it with synthetic perfumes and colors.

● **Garden chemicals**
If you garden organically, using only natural products on the garden, you will avoid causing any water contamination.

● **Do-it-yourself chemicals**
None of these should ever be poured down a drain or sink. Pure cellulose wallpaper paste can go on the compost heap: other chemicals should be poured into a deep hole well away from plants.

● **Car cleaning and maintenance**
The same rule applies here as for DIY chemicals – do not pour oil, battery acid, or car polishes down the drain. Do not use detergents in excessive quantities.

CAR
Car-cleaning detergents, polishes, and engine oil are all potentially toxic to river life. Washing them down a drain is the best way of ensuring that they do the maximum damage.

DISHWASHER
Dishwasher detergents are very caustic. They are a health hazard and in high concentrations they are toxic to the bacteria used in sewage treatment.

engineers know how to test for them and more or less how to eliminate them. Or most of them. But the by-products of modern farming methods are a different matter altogether. Run-off water containing biocides (p. 45) is causing serious concern for water supply engineers, principally because once biocides are in the water, they are almost impossible to detect or remove. But there is another agricultural menace to our water that may prove to be even more dangerous than biocides, and that is the so-called nitrate time bomb.

Nitrates are an important constituent of artificial fertilizer (the fertilizer that is applied to make up for our throwaway sewage systems). If you put nitrates on a growing crop, the plants will take up some of them, but they are never all absorbed in this way. Nitrates are very soluble and the first rain storm – or soaking from the irrigation spray – will wash them down into the soil. About half of the applied nitrate "leaches out" in this way.

Recent research has shown that it takes from twenty to forty years for this nitrate to seep downwards to the depth reached by borehole wells. Therefore the nitrate that farmers are putting on their soil now will not appear in drinking water for from twenty to forty years. Farmers did not *start* using excessive amounts of nitrate until about forty years ago – and they did not start using really massive amounts until the sixties and seventies, so we have yet to experience the full effect of this.

The chemical industry has put vast sums of its limitless money into trying to prove that nitrates are not harmful, and their many allies in exalted circles have aided them. But slowly, steadily, all over the world, the evidence builds up. High nitrate levels represent a very real threat to our health. In infants, high nitrate concentrations can impair the circulation to the brain, causing the "blue baby" syndrome. Over a longer period of time, nitrates may be a cancer risk: they are converted to nitrosamines in the body, and these are thought to be connected with stomach cancer.

COUNTERING THE NITRATE THREAT

The action that is currently being taken to defuse the nitrate time bomb is of the wrong sort. In West Germany, for example, road tankers are now

Water for irrigation Channel irrigation (*above*) and spray irrigation (*below*) soak land that is normally arid. As the irrigation water drains away, it carries with it much of the artificial fertilizer applied to the land. The result is water supplies polluted with nitrates.

transporting water to a number of villages in agricultural areas because the water supplies have been found unsafe to drink. A massive program of laying water mains from mountain areas to agricultural areas has been started for the same reason. But the nitrates are still in use.

In England many water supplies show nitrate concentrations well in excess of the 1 ounce per 164 gallons recommended as the upper limit by the EEC. The British have found a typically British way of getting round the problem. The government has simply told the EEC that it cannot meet the limit, and it has asked for a "derogation" to relax the standard to 1 ounce per 102 gallons. But, alas, many water authorities cannot even meet that standard and in many places it is 1 ounce per 80 gallons – *and it is going up all the time.*

The solution that we should obviously all be clamoring for is an end to a use of nitrates. But entrenched interests and conservative thinking are standing in the way of this. It is a fact that the manufacture of these substances is not necessary at all. Farmers grew perfectly good crops for thousands of years before these substances were invented. Their use has only become necessary because a sort of chemical rat race has developed, a race which no one is willing to stop. Farmers are egged on by the chemical industry to use larger and larger amounts of fertilizer on their crops. This further weakens the ability of the ground to produce large crops, so yet more fertilizer is needed to maintain high yields. The answer is always the same – keep taking the chemicals.

But there are – and always have been – farmers who are growing magnificent crops of every kind without a drop or a speck of artificial fertilizer. They are showing that it can be done. There is no need for anybody to starve. Nor to have their water poisoned. For the sake of drinkable water, at least, we should say "no nitrates", and persuade all our farmers to stop relying on purely chemical methods

Piped pollution If raw sewage contains only organic matter, it is quite harmless once it has decomposed. However, because most sewage also contains industrial effluent, it is a potentially hazardous substance.

of increasing production. This insidious menace can be stopped. Farmers must just get off the chemical treadmill before it is too late and they poison themselves and all the rest of us.

INDUSTRY'S LIQUID GARBAGE CAN

There has long been a description of people who like to kick up a row and prevent the world from becoming too sleepy a place. The term is "hell raisers". Is it not the time to coin another such term – but for a very different class of people – the "hell makers?" There is a growing army of people on this Earth now that, if given their own way, will turn this planet into a hell. The events at Chernobyl, the Russian nuclear plant, are part of this and are just one example of the direction in which this army is marching. Our planet – the air, the soil and the water – is becoming an enormous garbage can for the industries of the developed countries.

We can clean up what comes out of our own homes, and try to persuade farmers to cut down on nitrates, but the waterborne waste that comes out of factories is another problem altogether. Industry uses waterways as a cheap way of getting rid of waste. Every country has regulations to prevent industrial water pollution, but still it goes on. Industrial effluent often poisons waste treatment processes because they rely on bacteria, and these can be killed by some types of factory waste.

Throughout the industrialized world, there is increasing concern about the waterborne substances discharged by factories. Many of them end up in water supplies. In Britain 10 percent of the aquifers used for domestic supplies of water contain greater concentrations of chlorine-based solvents than recommended. In continental Europe and North America the situation is the same. Heavy metals (substances such as lead, mercury and cadmium), process chemicals such as metallic oxides and by-products from the pharmaceutical industry are steadily accumulating in the water cycle. As yet, we have no idea what long-term effects any of these will produce. We may not suffer from waterborne epidemics, but we may be storing up poisons which will have a devastating effect on our children.

The scale of industrial and agricultural water pollution is astounding. Three million tons of mercury are poured through our rivers each year – enough to poison the world's entire human population. Six million tons of oil also find their way to the sea. None of this reaches its final destination without trace amounts getting into drinking water.

A recent case in North America has shown how seriously many companies take water protection. The Hooker Chemical Company has an enormous chemical factory near Niagara Falls. In the 1940s it began to dump chemical wastes in an abandoned canal, which connected with the Niagara River. Subsequently this was earthed over and an elementary school built on it. When many people in the area became grossly ill, the cat got out of the bag. It was discovered what had been dumped there, and the area was evacuated and the school and houses pulled down. The story of Love Canal, for that was the name of the place, became infamous worldwide – in the United States the name Love Canal has become synonymous with pollution. But by then the company had started another dump at a place called Hyde Park, below the Falls. This new dump is about 2,600 feet from the Niagara River.

In the early 1970s Canadian biologists (Lake Ontario, into which the Niagara River runs, is half Canadian) began to notice that the gulls' eggs around the lake were not hatching. Tests found that they contained dioxin, a chlorine-containing chemical which is a by-product of pesticide manufacture. Direct discharges of dioxin into the river were stopped for a few years and the amount of dioxin in gulls' eggs declined. But then it began to rise again, and then it was realized that the chemical factories had started burying their wastes near the river and the chemicals were seeping through the earth and rock and getting into the water. The dump at Hyde Park contained at least *one ton* of dioxin, one of the most powerful poisons known to man. Canadian scientists reckon that 3 ounces mixed in with the vast volume of water in Lake Ontario would be enough to affect the gulls' eggs. And here, sitting in a shallow trench a few thousand feet from the river, is a *ton*!

However, the discoveries about Love Canal and Hyde Park were overshadowed by a new scandal. Divers working for the city of Niagara Falls found

that the water in a tunnel bringing drinking water from the river right by the Hooker factory had burned holes in their wet suits! The water was analyzed and was found to contain the same set of deadly chlorinated organic poisons that had been found at the Hyde Park dump. Investigations were made and the company was finally forced to admit that it had *another* dump there, in the factory grounds, that it had managed to keep quite secret.

The U.S. Environmental Protection Agency sued Hooker to force them to take remedial action. Hooker had plenty of money to fight the case (it was stated that the company had predicted sales of 1.7 billion dollars) and not surprisingly it won it, by arguing that, since there were so many other discharges of toxic chemicals into the river (including two of its own) how could it be claimed that any person had been harmed by leakage from that one particular dump?

And so the dioxin continues to seep into Lake Ontario and down the St. Lawrence River, destroying countless millions of living things and turning what was once a magnificent ecosystem into a sick and dying one. One could visit practically any country in the world and go to any industrial area of it and find similar situations. For example, analysts have noticed that the waters of the mighty River Rhine contain large amounts of phenol, hexachlorcyclohexane, hexachlorbenzol, pentachlorphenol, fluoride, mercury, nickel, zinc, copper, chromium, lead, manganese, arsenic, phosphates, ammonia and nitrates. And this is in a river where a clean up has been in progress. No wonder we see very little of the Rhine Maidens these days. Where the need to make a profit clashes with the need to keep water clean, profit making wins.

MAKING THE POLLUTER PAY

If the growth of this sort of pollution continues at its present rate, the world will be hopelessly contaminated not too far away in the future. Many companies assume that they have a right to use rivers as their dumping grounds. But we should be saying exactly the opposite: they have no right to do this, and if they do, they must put to rights any damage thereby caused. The rule should be "the polluter pays".

The dumping into the sewers and rivers of known poisons could be stopped immediately. Our governments should ban it. There are perfectly good ways (if rather expensive ones) of cleaning up industrial effluent. Settlement tanks and absorbing agents can remove many pollutants, while metal-tolerant bacteria can be used to concentrate lead, copper and cadmium so that it can be removed. In nearly every case, all that is needed is a financial or legal incentive to install the necessary equipment.

BACK TO THE SEA

The older among us were brought up to think of the boundless ocean as being the symbol of purity. Its extent was so vast that nothing, we thought, that mankind could do could pollute it or degrade it. In the last couple of decades we have had to revise our opinion. When Thor Heyerdahl drifted across the Atlantic in a reed boat he was appalled at the amount of rubbish he found right out in the desert wastes of the ocean. The banned pesticide DDT has been found in the fat of penguins in Antarctica (having worked its way right around the world in the marine food chain) and radioactivity from Sellafield has been detected as far away as the coast of Greenland.

The oceans have been found not to be, after all, boundless and mankind has been found quite capable of degrading them. The time has come when we have got to look to outer space for pristine purity: we have managed to foul up our own little planetary nest.

But it is the coastal and estuarine waters that are bearing the brunt of our impact most, and after them the comparatively shallow waters over the continental shelves. The deep ocean is still capable of absorbing much of our abuse. At least the criminal dumping of radioactive waste in the deep Atlantic has been stopped by the responsible action of the seamen who are refusing to dump it any more (although future generations will have to reap the bitter harvest sown by what has already been dumped). But alas the continued degradation of estuarine and coastal waters goes on unabated.

Partially enclosed seas like the Baltic, the Mediterranean, the North Sea and the Sea of Japan are already being grossly polluted. They have been used as dumping grounds for the disposal of every kind of filth we want to get rid of for too long.

We are land animals and have no right to destroy other environments. Estuaries are (or, perhaps more accurately, were) the most interesting and beautiful of places. Half-sea, half-land, invaded twice a day by the refreshing and renewing ocean and alternately twice a day by that other ocean – our atmosphere – they are inhabited by a whole

POSITIVE ACTION
Cleaning up coasts and estuaries

● **Keep a watch**
Many inshore pollution incidents are the result of illegal dumping. Water belongs to all of us: reporting offenders helps to prevent them from repeating their actions.

● **Boycott polluters**
A number of household products, particularly paints and plastics, can cause severe water pollution during manufacture. Although pollution may be within legal levels, it does not have to happen at all. It can be stopped by consumers boycotting persistent polluters.

● **Prevent pollution afloat**
To protect the marine environment, boat owners should dispose of garbage on land. We do not have a right to pass on our garbage to marine life.

● **Action on sewage**
If you live in an area where sewage is disposed of by dumping it at sea, let the authorities in charge of waste disposal know that they should use land-based methods of disposal instead.

● **Do not use antifouling paints**
These toxic paints, which are used to prevent encrustation of boat hulls, kill shellfish not only on boats but also around them. Responsible boat-owners should not use them.

● **Watch out for plastic hazards**
Clear plastic is particularly dangerous to aquatic animals because it is invisible underwater. Fishing lines and the plastic retainers from packs of cans can strangle birds and seals (see p. 86). Picking it up will prevent this happening.

Poison underwater
An outfall pipe (*above*) discharges pollutants into shallow water – killing marine life.

Cancerous fish
Surveys in the North Sea show that half of some flatfish species (*below*) have pollution-induced cancer.

fertilized farmland, the "soft-shelled crabs" are becoming creatures of the past and the same fate is happening to all the other fisheries in that great tidal inlet. The large fishing community there is going the way of the oysters and crabs.

The same sad theme is repeated with variations all around the world. In Britain, the Mersey is dead and stinking and the Humber is not much better. This damage is being done by us. Each one of us is personally responsible for it, and responsible to see that it stops. Even a small degree of responsibility should urge us to support such organizations as Greenpeace which is fighting hard to stop such abuse, and it should also cause us to press politically for a civilized sewage policy – which means putting human waste back on the land, not in the nearest piece of sea. We should lobby for much higher fines for oil polluters and we should give our votes to any political party that gives a high priority to stopping pollution. We should personally – and boldly – speak up and protest when we find polluters at work. "Polluters at Work" – what about that painted on all factories and other establishments that are guilty? They are only doing it to make money for people who already have too much.

We often tell ourselves "this is our world". Well, it is not just our world. It is also the world of millions of species of creatures which have just as much right to it as we have, whether they be great or small, and live by land or by sea or out in the vastness of the great oceans.

world of creatures amazingly adapted for life in what is a very difficult environment. Further, from the selfish human point of view, estuaries – and also deltaic waters – can be highly productive places. The vast shoals of sardines once found off the Nile delta went far to satisfying the protein needs of the Egyptian people – before the Aswan Dam cut off their natural food supply, the mud from the Nile, so that the sardines have now disappeared. Such great sea inlets as Chesapeake Bay are dying rapidly, owing to run-off from over-

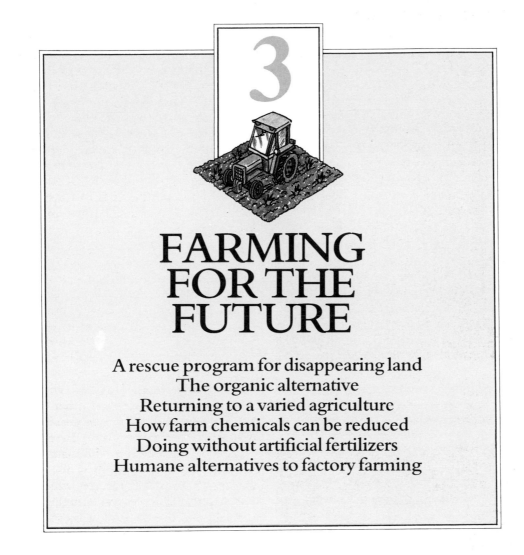

3

FARMING FOR THE FUTURE

A rescue program for disappearing land
The organic alternative
Returning to a varied agriculture
How farm chemicals can be reduced
Doing without artificial fertilizers
Humane alternatives to factory farming

If you were to ask any farmer, fifty years ago, what kind of farm he ran, the answer would probably have been a "mixed" one. By that he would have meant he produced a bit of this and a bit of that: wheat, barley, beans perhaps, potatoes, cows, pigs, goats, sheep (if he had the room), ducks, geese, and almost certainly hens, because they can pick up a living on next to nothing. The average prewar farm was a general business, and the people that worked it were all-arounders.

Now that sort of farm has rapidly become a thing of the past. Today's farmers are not all-arounders but specialists, technicians skilled at raising just one or two kinds of crop or animal, and very little else. They do this so single-mindedly that they turn out their produce in great quantities – so much so that often we cannot eat our way through it fast enough, and it piles up in wheat and beef "mountains".

To some people, this is the sort of progress that we should all be thoroughly pleased with. But industrial farming, or "agribusiness", is sowing a wind that our children will reap as a whirlwind. It is certainly producing food in enormous quantities, but it is doing so in a way that cannot possibly be maintained, and which (if we do not take action to stop it) will ruin the land forever. Modern farmers may produce grain and meat at knock-down prices at the moment, but this has to be paid for dearly by the land they work. Looked at in the round, today's farming is at best a loss maker, and at worst an ecological disaster, and something must be done about it.

THE DISAPPEARING LAND

Soil erosion is the most serious threat to our planet at the present time. It is also one of the least noticed, probably because of its undramatic and gradual nature. Precious topsoil is being lost to world farming at the rate of 25 billion tons every year, that is about 7 percent of the world's soil every decade. Without soil this planet would be as lifeless as the Moon.

Although American farmers are among the most productive in the world, their land is disappearing beneath their plows. For a number of decades now, they have produced grain in ever-increasing amounts, but financial constraints mean they have to do this with a minimum of human labor. The consequence is vast expenditure on mechanization and chemicals, and a system of farming that has disastrous side effects.

Under normal circumstances, erosion is largely balanced by the formation of new soil from weathered rock. Natural vegetation protects this soil once it has been created. But today's farmers work the ground so hard that the soil is blown or washed away faster than new soil is created to replace it. Since 1977, 1.7 billion tons of American farmland have disappeared *every year*. That means that for every ton of corn that the farmers produced, over five hundred tons of soil are gone for good.

The present rate of loss is simply not sustainable. The soil of the eastern half of the Great Plains of North America is going down the Mississippi to the sea, while much of the soil of the western, drier, areas has already blown away during the terrible "Dust Bowl" years in the 1930s. There are strong fears that the rest will follow shortly.

The situation in Europe is not as bad, but the warning signs are there. All along Europe's coastlines, rich brown estuaries are pouring good soil into the sea, from where it will never return. And when any of this soil is washed away, centuries of patient and careful cultivation go with it.

DOWN ON THE FACTORY FARM
How modern farming damages the land

Today's farms have little use for the skilled husbandry, which was once the guiding principle of working the land. The emphasis today is solely on productivity – high input in exchange for high returns. But in assessing the returns, two important considerations, what happens to the land and the food it produces, are overlooked.

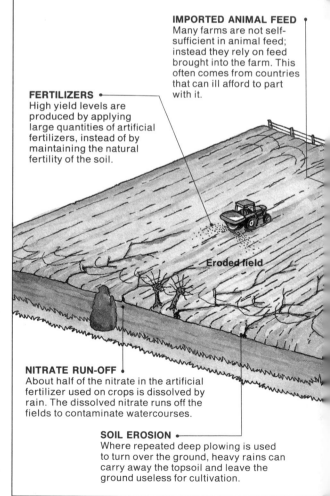

IMPORTED ANIMAL FEED
Many farms are not self-sufficient in animal feed; instead they rely on feed brought into the farm. This often comes from countries that can ill afford to part with it.

FERTILIZERS
High yield levels are produced by applying large quantities of artificial fertilizers, instead of by maintaining the natural fertility of the soil.

Eroded field

NITRATE RUN-OFF
About half of the nitrate in the artificial fertilizer used on crops is dissolved by rain. The dissolved nitrate runs off the fields to contaminate watercourses.

SOIL EROSION
Where repeated deep plowing is used to turn over the ground, heavy rains can carry away the topsoil and leave the ground useless for cultivation.

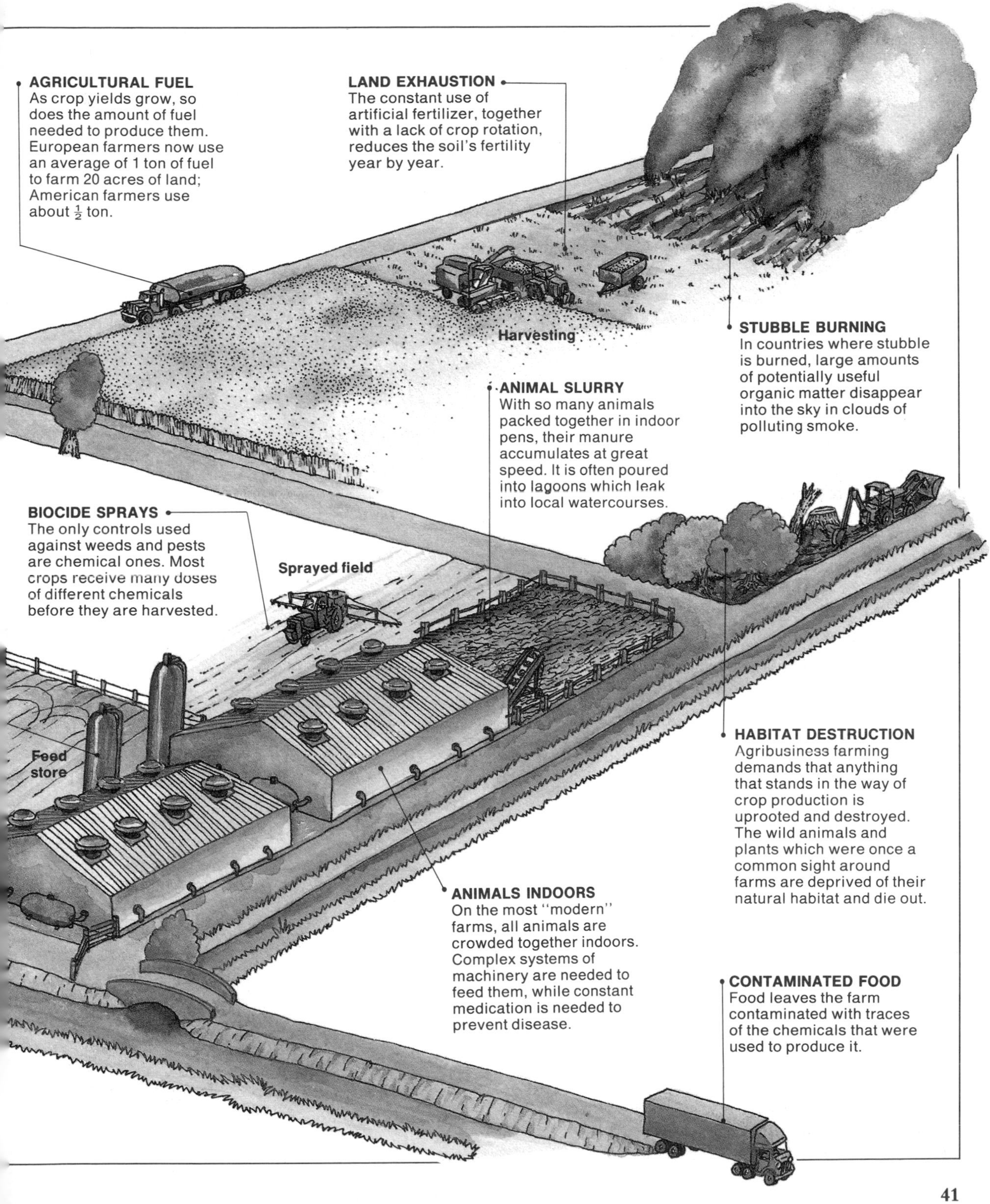

AGRICULTURAL FUEL
As crop yields grow, so does the amount of fuel needed to produce them. European farmers now use an average of 1 ton of fuel to farm 20 acres of land; American farmers use about $\frac{1}{2}$ ton.

LAND EXHAUSTION
The constant use of artificial fertilizer, together with a lack of crop rotation, reduces the soil's fertility year by year.

STUBBLE BURNING
In countries where stubble is burned, large amounts of potentially useful organic matter disappear into the sky in clouds of polluting smoke.

Harvesting

ANIMAL SLURRY
With so many animals packed together in indoor pens, their manure accumulates at great speed. It is often poured into lagoons which leak into local watercourses.

BIOCIDE SPRAYS
The only controls used against weeds and pests are chemical ones. Most crops receive many doses of different chemicals before they are harvested.

Sprayed field

HABITAT DESTRUCTION
Agribusiness farming demands that anything that stands in the way of crop production is uprooted and destroyed. The wild animals and plants which were once a common sight around farms are deprived of their natural habitat and die out.

Feed store

ANIMALS INDOORS
On the most "modern" farms, all animals are crowded together indoors. Complex systems of machinery are needed to feed them, while constant medication is needed to prevent disease.

CONTAMINATED FOOD
Food leaves the farm contaminated with traces of the chemicals that were used to produce it.

WHAT CAN BE DONE ABOUT SOIL EROSION?

What can we, sitting at our suburban breakfast tables in European or North American cities, possibly do to stop all this? The answer is simple.

Not all that long ago our forebears had to put in many hours of sweated labor to earn their daily bread. The result was that they valued it, and none was wasted. Now the average office worker can earn enough to keep body and soul together not in hours but in minutes, and as a result food is treated as a throwaway resource. We waste it horribly, and every year vast amounts of perfectly good grain, fruit and vegetables are deliberately destroyed in blind obedience to "market forces". No wonder modern farmers have come to treat their land in the same throwaway manner.

To put a stop to this appalling prodigality, and its tragic effect on the soil, we must put a far higher value on foodstuffs that are produced with some thought for the land. This means putting quality before quantity for once, and using quite different methods of farming.

THE ORGANIC ALTERNATIVE

Organic farming is a sustainable, biological approach to farming, one which does not treat land like an expendable commodity. There is little erosion on organic farms. The reason that there is little erosion is that the organic farmer operates a "closed" farming system. That means not buying in great quantities of chemicals and not relying on the constant use of machinery, but instead making use of the natural resources of the land and recycling all its nutrients.

The organic farmer is entirely dependent on maintaining a high humus content in his soil if he wants to get any sort of a crop at all, and humus-rich soil does not erode. Humus is decaying or decayed organic matter, that is matter with an animal or vegetable origin. It sticks the soil particles together and prevents them from either blowing or washing away. The soils of the Great Plains did not begin to blow away until the great store of humus that had been accumulated by ten thousand years of prairie grasses living and dying on it, and all the buffalo dung that went with that store, had been used up completely by "extractive" rather than closed farming. Farmers made no attempt to put back anything to replace what they had taken out. Then, when all the humus had gone, there was nothing to hold the soil together and it blew away. And that was the Dust Bowl – together with acid rain, perhaps the greatest cataclysm mankind has managed to bring about on this Earth so far.

Danger – erosion in progress
Soil erosion is not only caused by intensive crop cultivation: it can also be triggered by overstocking land with animals. A given piece of land can only support a certain number of animals. If this number is exceeded, the animals begin to destroy plants through their browsing and grazing. When this happens, the plant cover that protects the soil disappears, and the land becomes unstable. Erosion is the result.

HOW ORGANIC FARMING WORKS

Organic farming is based on the lessons learned in countless generations of working on the land. Farmers learned two things at an early stage. First, the importance of variety. If they grew just one species of crop on the same land year after year, like farmers do now, they ran into trouble. Insect pests, for example, profited from this reliable abundance of their favorite food and, as a direct result, multiplied rapidly and attacked the crops with vigor. Animals, like crops, did not thrive if too closely concentrated in single-species populations (as they are on farms today). It gave the parasites and disease-causing organisms an unfair advantage, and they became too numerous for the animals to cope with. Without the attentions of a pharmaceutical industry to keep them going, the infected animals would perish.

So the first lesson was "interfere too drastically with the natural balance and trouble will come of it." Farmers learned to rotate crops, never growing one species of annual plant year after year on the same land. They also learned the advantage of mixed stocking (keeping more than one species of animal) – the advantage being that one species of animal harmlessly ingests the parasites voided by another. If a sheep, for example, eats worm eggs voided by a horse the eggs will die, doing no harm to the sheep.

Second, they learned the importance of maintaining healthy soil. It was not enough to keep different types of animals, or grow different types of crops, if the soil was impoverished. The organic content of the soil was crucial. Farmers placed great importance on keeping the land in "good heart" as it was called, rich in organic material. All plant residues were returned to the land, generally carefully composted by mixture with animal manures first. In the "open field" system that survived in Europe for fifteen hundred years without causing any erosion, grazing animals were turned out on the stubbles after the corn harvest, and the dung of any housed or yarded animals was assiduously returned to the land.

Today's organic farmers use many of these traditional techniques to produce their crops and so avoid many of the problems that confront their big-business counterparts. Through the use of crop rotation, natural fertilizers, soil-improving plants, and the avoidance of agrochemicals, organic farming constantly adds to the soil. The organic farmer certainly uses powered machinery, but only where it is essential. No chemicals, less oil and an improving soil all add up to one thing – a cheaper and more natural system of production.

THE DANGERS OF MONOCULTURE

In 1970 a fungus that lives on corn, *Helminthosporium maydis*, ran rampant through the U.S. corn belt. It reduced the corn crop by 15 percent, and forced almost every farmer to switch his seed corn, to prevent certain disaster the following year.

The fungus did so handsomely because the same variety of wheat was growing right across the country as a "monoculture". The fungus's food was almost unlimited. This experience has been repeated many times where farmers have put all their faith in just one variety of a plant or animal. Sooner or later, disease breaks out, and spreads like a high-summer fire because there are no different varieties to check its progress.

The preservation and use of old breeds – be they plant or animal – are tremendously important. At present, with the passing of every year, dozens of

---

POSITIVE ACTION
How to help the preservation of rare breeds

- **Buy local varieties**
  Buying local varieties of fruit and vegetables will help to prevent them being replaced by their standardized big-business counterparts.

- **Use a seed bank**
  Some organic farming and gardening associations maintain seed banks which preserve old varieties of cultivated plants for the future. Joining an association will allow you to grow and exchange unusual and valuable seed. For addresses, see p. 185.

- **Rare breeds preservation societies**
  A number of organizations now exist to safeguard the future of rare farm animal breeds. For addresses, see p. 185.

old crop and stock varieties die out just because they don't conform to some bureaucratic standard. Gone are the old long-stemmed wheats, the tall-trunked apples, the mongrel hens and the spotted pigs. In their place we have regulation size and shape, and less and less variety in everything we buy.

Relying on just a handful of species for your food is highly dangerous. Relying on just one or two varieties of each species is more foolhardy still. So, wherever old varieties still survive, they must be kept going – this applies even to the ancient apple tree at the bottom of the garden. They are our genetic insurance against pests and disease.

THE AGROCHEMICAL ADDICTION

If creeping soil erosion is the greatest threat to all land life on this planet, galloping chemicalization runs it a good second. Apologists for the chemical industry are constantly reminding us that all matter is made of chemicals. As long ago as 1846, the London *Quarterly Review* stated that a man is "forty-five pounds of carbon and nitrogen diffused through five-and-a-half buckets of water". Maybe we nonchemists may be forgiven for believing that Shakespeare, to take one example, was more than just that?

In the widest sense of course all matter *is* chemical: a farmer who puts good manure on his land is applying chemicals just as much as if he were spraying it with 2,4,5-trichlorophenoxyacetic acid. But there is a difference between substances that occur naturally and substances that are produced by industry.

Every living creature on this planet, animal or vegetable, has evolved by natural selection to exist in a certain chemical environment. Terrestrial organisms have evolved to live in an atmosphere made of a mixture of oxygen and nitrogen, with a small proportion of carbon dioxide, and traces of a few other gases. They have not evolved to cope with the ever-growing battery of synthetic chemicals now sprayed on them. It is these chemicals which make today's farming so different from anything which has gone before it.

Agricultural chemicals fall into two general categories – biocides and fertilizers. "Biocide" is a

polite word that has come to replace the old English word "poison", which doesn't sound so nice. A biocide is a substance that destroys life, usually (but not always) that life that reduces crop yields. Fertilizers on the other hand are chemical nutrients that enable plants to grow more vigorously in poorly managed soil. Both are now used in prodigious quantities.

THE SIDE EFFECTS OF BIOCIDES

The skill of the modern chemist has been to make chemicals which destroy some forms of life without completely destroying others. It's not an easy job. Agricultural laboratories have the ability to turn out thousands if not millions of chemical substances, but only occasionally one happens to have a lethal effect on particular farmland pests.

Assessing the effect of a potential biocide can be extremely difficult. It doesn't take long to find out if a chemical kills aphids, for example, but it is much harder and more time consuming to determine what it will do to the birds that eat the dying aphids, or the young of those birds, or the animals that feed on them. All animals and plants are part of food chains of immense complexity, and it is virtually impossible to poison one organism without sooner or later having an effect on others. The full consequences of using a particular chemical may take years to show up – much longer than any laboratory test.

There is a particular problem in making sure that biocides end up where they are intended to go. You can design a chemical that attacks fungus on wheat, test it in a wheat field, and produce apparently satisfactory results. But you cannot reproduce all the conditions under which that chemical will be used. You cannot, for example, test it by spraying it so carelessly that it blows far away and in through the windows of nearby houses. You cannot spray it onto passing cars, people or animals. You cannot spray it when the weather is wrong, and then harvest the wheat at the wrong time so that you can see what it does to the people who eat it. But this is what happens every day on farms and the land around them.

There are countless examples of biocides creating havoc in the natural chain of eater and eaten. DDT,

THE BIOCIDE EXPLOSION

Over the last twenty years, the production of biocides or agricultural poisons has been one of the largest growth areas in the chemical industry, and today few farms are without their store of poisons waiting to be put on the land. This table shows what the main types of biocide are and what effects they have. It also shows what their estimated production levels are over nearly two decades. Although the three types of biocide described here make up the bulk of those applied to crops, even more chemicals may be used to combat crops suffering from less common pests such as nematode worms and mites.

FUNGICIDES

What they kill
Fungicides are used to kill the often microscopic fungi that infect growing crops, fruit and stored seed. These fungi include the mildews, rusts, pin-molds, and yeasts.

What they contain
Most fungicides are based on compounds containing metals such as copper and sometimes mercury, or on hydrocarbons containing sulfur.

Health hazards
Fungicides are often sprayed directly onto the part of a crop that is destined to be eaten. Traces of fungicides are frequently found on fruit and vegetables and these can build up within the body with as yet unknown consequences.

WORLDWIDE USE
The use of fungicides will have grown by nearly 20 times in the same number of years.

(millions of tons)

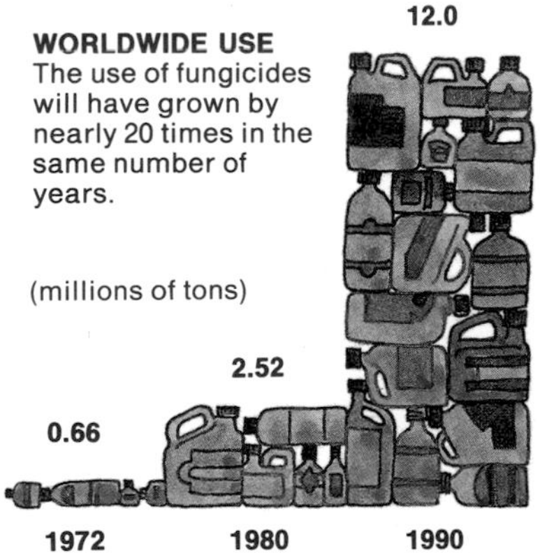

12.0

2.52

0.66

1972   1980   1990

INSECTICIDES

What they kill
Insecticides kill aphids, weevils, and other insect pests, and are used on growing crops, and to a lesser extent on stored grain. As well as killing pests, they often also kill the insect predators which feed on them.

What they contain
The most commonly-used insect-icides are the organophosphates (compounds containing phosphorus) and chlorinated hydrocarbons (compounds containing chlorine).

Health hazards
Many insecticides have proved very persistent. They do not degrade rapidly and can be passed on in food to cause conditions such as liver failure.

18.0

WORLDWIDE USE
The use of insecticides will have increased by nearly 17 times in two decades.

4.8

1.08

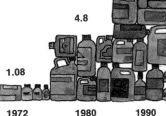

1972   1980   1990

HERBICIDES

What they kill
Herbicides kill plants. They may be non-specific (killing all plants when land is cleared) or they may be specific, for example killing broad-leaved weeds growing in cereal crops.

What they contain
Herbicides are a group of highly varied chemicals. Many are poisonous mimics of natural substances within plants. Once they are absorbed, they kill plants by blocking their metabolism.

Health hazards
Some herbicides are deadly if accidentally consumed. Others may cause nonfatal illnesses when eaten in food.

16.0

WORLDWIDE USE
The use of herbicides will have grown by about 15 times in two decades.

4.08

1.08

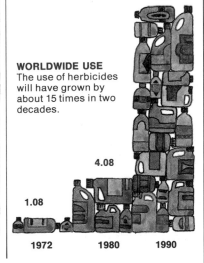

1972   1980   1990

for example, an organochlorine insecticide originally pronounced safe, was banned in the early 1970s
in most countries (although many of them, rather
hypocritically, still make it for export). DDT was
marvelous stuff for killing insects. However, it also
killed birds by reducing the thickness of their eggshells which made it impossible for them to stay
intact during incubation. But worse still, DDT
showed toxic effects on humans – all this in a
chemical that was hailed as completely safe.
Hundreds of other biocides have had a similar
history – declared safe, marketed with great
enthusiasm, and subsequently found to be
dangerous.

THE BIOCIDE BUILD-UP

Many biocides are "systemics" – they enter into
the cells of a plant either to protect it from fungi or
attack by insects. They may remain in the plant for
weeks or months. When we eat the plant we will
not die – the amount of poison is too small – but it
will not do us any good. The poisoning goes on
continually, and much of it eventually percolates
down to the dinner table. No one yet knows exactly what cumulative effect these chemicals will have.

From 1950 to 1967 pesticide use in the United
States shot up by 168 percent, a threefold increase,
and now most wheat in both North America and
Europe is sprayed at least six times before
harvesting. Some crops, like cotton and the field
pea, fare even worse.

Despite this heavier chemical soaking, the pests
are actually increasing. Since 1960, the number of
potentially damaging insect species resistant to
pesticides has increased from 160 to about 450. In
some parts of the United States, farmers have been
allowed to resort to banned pesticides because the
infamous Colorado potato beetle has developed a
resistance to newer chemicals. Similarly, in spite of
constant dosing with fungicides, fungus diseases in
plants are getting worse. A constant stream of new
diseases afflicts the major food and fiber crops –
diseases that had never been heard of fifty years ago.

To combat the problem of pest increase, chemical industrialists and agricultural advisory services
always come up with the same solution – use *more*
chemicals. "Don't worry," they say. "Whenever

the disease-causing organisms take a jump ahead
our researchers will invent yet more virulent
mixtures to do them down. Keep on spraying. If
nine biocide applications a year aren't enough to
keep your wheat disease- and weed-free, try ten!"
And the pity of it is they do.

WHY THE BIOCIDE TREADMILL KEEPS TURNING

Chemical companies have an obvious vested interest
in keeping biocide sales up. The clash of public and
private interests that this creates was well illustrated
in the United States aldrin/dieldrin debate in the
early seventies. Both these substances were
suspected of being carcinogenic. However, when
the newly-formed U.S. Environmental Protection
Agency put forward measures to prohibit their use,
Shell (the manufacturers) brought a huge battery of
high-powered artillery to bear in contesting the
proposals. Aldrin and dieldrin were under suspicion
from 1962 after a Food and Drugs Administration
report linked them with cancer. It was *fourteen* years
later that the ban came into force, and this happened
in a country which leads the world in safety.

In 1986 the main federal law governing biocide
use was amended so that each new product was subject to a registration fee of $150,000. This money
will be used to monitor the manufacturers' safety
tests, but the size of the fee is unlikely to reduce
the line of products awaiting certification – the
money to be made from selling biocides is on
another plane entirely.

The chemical industry is globally organized.
You can see this for yourself. Simply buy a farming
magazine, and flip through its pages. You will find
that nearly all the advertisements – and certainly all
the expensive advertisements – have been placed by
the chemical industry. Search through the editorial
material and notice how many items there are
which are critical of the chemical industry. Very
few if any. A substantial part of the revenue of the
agricultural press comes directly from the chemical
industry, so it would be rather strange if there were.

So, despite the tragedies of Bhopal and Seveso
(both factories which were involved in the production of biocides), chemical plants continue to
pump out their poisons onto our planet.

A CHEMICAL TIMETABLE

During its life in the field, the humble pea comes in for more chemical attention than many people would receive from a lifetime's visits to the doctor. Growers of crops such as this adhere to a rigid program of spraying to prevent pests or weeds gaining a toehold.

Here the total number of chemicals used on a pea plant are shown at different stages in its life. By the time it is harvested, it may have been subjected to a total of 10 applications of herbicides, insecticides, and fungicides.

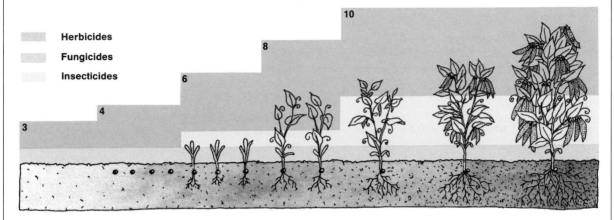

Herbicides

Fungicides

Insecticides

PRE-GERMINATION
The ground is kept weed-free with two different types of herbicide. The seed itself is often "dressed" with a fungicide to prevent mildew.

GERMINATION
As soon as the plant begins to grow, an insecticide is applied to kill weevils. More herbicides are used to control germinating weeds.

FLOWERING
Aphids, pea moths, and pea midges call for a number of extra doses of different insecticides. More herbicide is still applied.

POD FORMATION
The aphid, moth, and midge insecticides are sprayed on regularly as the pods ripen. By this stage up to 10 separate treatments have been applied.

HOW BIOCIDE LEVELS CAN BE REDUCED

We are creatures of the soil and we poison the soil at our peril. The spade's-depth of tilth on the surface of the land is all there is between us and starvation, so we should treasure and maintain the chemical well-being of the soil above all other things – even money. This means cutting right down on agricultural poisons.

There is one change that any farmer can put into effect immediately without any loss of crops. That is giving up "cosmetic" and "calendar" spraying. With cosmetic spraying, biocides are used simply to improve the appearance of a crop by removing the odd weed that is actually doing little harm. This cosmetic spraying causes quite unnecessary pollution. Calendar spraying means spraying according to a fixed schedule, regardless of what is happening on the ground. Again, it often results in biocides being used to counter nonexistent threats.

POSITIVE ACTION

How to reduce the use of biocides

- **Buy organically grown food**
  Because organic farmers do not use biocides, buying their produce will ensure that an increasing area of land is farmed without the use of chemicals (for information on wholefood, see p. 70).

- **Buy fresh produce**
  Significantly more chemicals are used on food crops which are destined to be packed and sold far away from their place of origin. Although fresh produce may take more time to prepare, it does not encourage the use of biocides.

- **Ban biocides at home**
  It is no good insisting that farmers should not use chemicals if they are still used in and around the home. Avoiding the use of garden biocides is an important part of reducing the general level of these poisons.

Of the more permanent ways of shedding the soil's biocide burden, the first method is called Integrated Pest Control, and is a very effective mixed approach. IPC is a method of controlling pests of both crops and stock (not eliminating them – you can never eliminate them) by intelligence and common sense – not just by blanketing the land with poison. It is proving highly successful. Over a few years it has, for example, increased the apple harvest in Washington and the citrus crop in California. With IPC, biologists study a problem in depth and then devise means of mitigating pest damage by such things as encouraging predators (other creatures that prey on harmful insects, for example), rotating crops, varying planting times, and then carrying out very limited and closely targeted spraying when all other methods fail.

With IPC, it is the combination of measures which does the trick. Instead of relying on one kind of remedy, the problem is attacked on a broad front. IPC not only deals with specific problem pests, it makes life more difficult for all of them, and it is more likely to stay effective than biocides alone.

Organic farming, the second method, rejects the use of biocides altogether. The organic farmer accepts that pests will take their share of any crop, but by rotating crops and weeding, keeps that share to a minimum. The soil is entirely free of synthetic chemicals, although in land "converted" from industrial agriculture, the complete cleansing process may take a number of years.

THE RISE OF ARTIFICIAL FERTILIZERS

The Industrial Revolution might have been fueled on coal but it was fed on Chilean bird guano.

In the nineteenth century vast deposits of guano (the accumulation of thousands of years of seabird droppings) were discovered on the desert shores of South America and of Southwest Africa. Guano is a natural fertilizer, rich in nitrogen and phosphorus, two elements vital for plant growth. For half a century a mighty fleet of sailing vessels carried the stuff around Cape Horn and to Europe or North America where it was put upon the land. It increased productivity enormously. But eventually the guano ran out, and farmers were forced to look for substitutes for this wonder substance.

Although nitrogen makes up over three-quarters of the Earth's atmosphere, plants cannot use it directly. They need it in the form of nitrate, that is, nitrogen that is combined or "fixed" with oxygen. On land the only organisms capable of carrying out this process are bacteria.

Nitrogen-fixing bacteria live both freely in the soil and within the roots of plants of the pea family – not only peas themselves but also beans, alfalfa, lucerne, clover, lupins and so on. These "leguminous" plants, as they are known, contain colonies of nitrogen-fixing bacteria within nodules on their roots.

In well-tilled soil, the free-living bacteria and the bacteria housed in the root nodules of leguminous plants together provide a constant supply of nitrate for plants to use. The trouble is that they do not do it fast enough for modern farmers. So in a misguided attempt to improve on nature, farmers buy artificially fixed nitrates in the form of commercial fertilizers.

A CASE OF DIMINISHING RETURNS

The fertilizer bag has become the hallmark of today's agriculture, and making what goes in it is one of the world's greatest industries. The production of nitrogen-containing fertilizer is carried out on a huge scale by passing air and hydrogen over a catalyst. The hydrogen is nearly always derived from natural gas, and so in energy terms, it is a very expensive process.

Through the use of this artificial fertilizer and genetically improved crop varieties, farmers can do without good husbandry. They can grow record crops simply by dumping artificial fertilizer on the land. But they are, in effect, merely buying today's readily-available fossil fuel energy (which makes the fertilizers in the first place) and turning this energy into food. Because it is a noncyclical process it cannot be self-sustaining.

At the end of World War II the average corn output of farms in Illinois was 2,492 pounds per acre. Twenty years later the same farms were averaging 4,717 pounds per acre. A magnificent improvement, you might think. It was. But during the same period those farmers went from using 10,000 tons of artificial fertilizer to 400,000 tons of it. So in Illinois alone there was a forty-fold increase of fertilizer consumption to produce slightly less than a doubling of crop yield, and bear in mind that some of this extra yield was due to improved crop varieties, not chemicals.

These figures are not peculiar to Illinois either. In 1970, farmers in Europe and the United States each got through about 17 billion tons of artificial fertilizer. By 1983, the figure for both groups had risen to 22 billion tons. To give an idea of what this represents on the ground, Dutch farmers, who lead the world in fertilizer use, now apply one-third ton of artificial fertilizer to each acre of their land every year.

Using artificial fertilizer is like getting more warmth from a fire by emptying the coal scuttle on it: the fire blazes brightly before burning itself out. In the United States a quarter of the energy from a limited supply of natural gas now goes to making fertilizer, and a whole generation of farmers has become completely hooked on it.

DOING WITHOUT ARTIFICIAL FERTILIZERS

What would happen if farmers gradually abandoned artificial fertilizers? According to the fertilizer industry the result would be catastrophic.

This, of course, is rubbish. But one thing is sure, our farmers will have to give them up, so the sooner the better. Agribusiness farming is entirely dependent on the planet's supplies of gas and oil. Without the fertilizer these fuels produce, industrial farmers could not raise a single harvest. When this fuel becomes too expensive to use (as it will when nearly all the oil and gas has been squandered) the supply of artificial fertilizer will dry up.

The economics of using artificial fertilizers are

Record harvests – at a price
Each trailer-load of grain (*above*) delivered by agribusiness farming requires, in its production, up to one-tenth of its own weight of artificial fertilizer.

High-input farming
The large quantity of fuel consumed by farm machinery (*left*) is one more cost incurred by high-input intensive farming.

not as persuasive as their makers would like us to think. Practically all farmers in the world are heavily in debt. They must make enough money this year to pay the interest on mortgages and loans, and the only way to do this is to continue dosing the land with heavier and heavier supplies of fertilizers to try to produce the greatest possible tonnage of saleable crop. Farmers cannot afford to give their land a chance to recover its fertility between crops. They cannot afford to pay for the labor that would enable cattle to be brought to the farms, and which would carry and spread the resultant manure. So they are driven deeper and deeper into debt to pay for expensive machines and the increasingly expensive fertilizers needed to grow the crops to pay the interest on loans.

According to the Department of Agriculture nearly one hundred thousand American farmers face technical insolvency: if the banks foreclosed on them they would have to sell. Many of them are actually doing so.

What happens when a farmer gives up artificial fertilizers and "goes organic"? This is the question that is being addressed at an experimental farm in Nagele, Holland. The answer is that, although the crop yields are slightly lower than on a "normal" farm, the costs prove to be lower still. Moreover, a survey of existing organic farms in Switzerland has shown that the net income for these was the same as for their agribusiness counterparts.

Once farmers have made the transition, which may take four or five years, they find that financial profits actually go up. Organic farming *does* pay. Their production per unit area will almost certainly go down (from ten to twenty percent is the common experience) but because they no longer have to lay out huge sums on chemicals they make more money. As chemicals become ever more expensive this differential will change increasingly in favor of the organic farmer.

But how would a drop in production affect the world as a whole? If 10 to 20 percent less wheat is grown, some of this reduction would simply remove the stockpiles that are expensively stored as grain mountains. It is true that there would be less to export or give away to countries that need it. But dumping huge quantities of free food on needy

WHY ORGANIC FARMS MAKE SENSE

Organic farming is a practical demonstration of the fact that a simple approach is usually a better one. Farming the land without an armory of chemicals, and instead using natural techniques and materials, produces reliable crops that do not carry with them the danger of contamination by farm chemicals. The key to organic farming is scrupulous attention to the well-being of the soil, an art which has been brushed aside by today's big-business farmers.

POSITIVE ACTION
The principles of organic farming

The aims of the organic farmer are summed up in the six basic standards below. These have been drawn up by the International Federation of Organic Agriculture Movements, an organization that promotes organic farming.

- **Localism**
  As far as possible, an organic farm is run within a closed system, drawing upon local resources instead of relying on raw materials from outside.

- **Soil improvement**
  Instead of depleting the soil, the organic farmer aims to maintain and improve its natural fertility. This rules out the use of artificial fertilizers, as these do not improve the soil over the long term.

- **Pollution abatement**
  The organic farmer takes steps to avoid all forms of pollution when raising and harvesting crops. This excludes the use of all synthetic biocides, as these pollute ground, wildlife and food.

- **Quality of produce**
  As well as producing food in quantity, the organic farmer places an emphasis on a high nutritional quality in the farm's output.

- **Energy use**
  On an organic farm, the use of fossil fuels like oil is kept to a minimum.

- **Employment**
  The organic farmer aims to produce employment which is both satisfying and financially rewarding for farm workers.

THE AGRIBUSINESS FIELD'S YEAR

The high yields of an agribusiness field are only achieved by repeated chemical assistance in the form of fertilizers and biocides. The natural life of the soil is so much reduced that it cannot maintain fertility, and each successive harvest leaves the soil in a worse state than it began.

THE ORGANIC FIELD'S YEAR

In an organic field, only natural materials are applied to the land, and as a result the soil is rich and fertile. The field will produce a good crop every year with no chemical input. Instead of being impoverished, it is actually improved by the annual farming cycle.

SPRING

Above ground
Before and during the emergence of the seedlings, herbicides are used to reduce weeds. Granular fertilizer is applied to artificially boost the vigor of the crop.

Below ground
Because no manure has been applied to the land the previous year, the soil is compacted and lacks organic matter.

Above ground
As the seedlings appear weeds are controlled by repeated hoeing, which also opens up the ground. No biocides are used.

Below ground
The soil is rich in organic matter produced by compost or manure applied the previous winter. It has an open structure and is rich in earthworms.

SUMMER

Above ground
As the crop starts to produce ears, regular spraying is used to keep down insect pests. All weeds have been killed by the application of herbicides.

Below ground
The fertilizer is dissolved by rain and carried below ground. Only a proportion is taken up by the crop.

Above ground
At this stage all that distinguishes the crop from its agribusiness counterpart is a scattering of weeds.

Below ground
Nitrogen produced by the decay of compost and manure is taken up by the growing crop. The roots of the crop spread easily through the soil.

AUTUMN

Above ground
After repeated biocide spraying, the crop is harvested. The straw is often burned off, wasting organic matter.

Below ground
The dissolved fertilizer not taken up by the crop is carried further underground.

Above ground
After harvesting, the land may be left for many months before ploughing. Nitrogen-producing wild plants, like clover, which were hidden in the crop, now spread over the ground.

Below ground
The decaying compost and manure continue to add to the soil's fertility.

WINTER

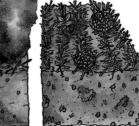
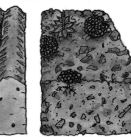

Above ground
As soon as the crop has been harvested, the remains of the stubble are ploughed back into the soil. The same crop is then immediately replanted.

Below ground
All the fertilizer has been washed away to contaminate the water cycle, and the soil is left in a worse condition than it began the year.

Above ground
Compost and manure are scattered over the field. A covering of vegetation protects the field from erosion until it is ploughed for the next year's crop.

Below ground
During the winter months, the processes of decay slow down. The year's crop has not reduced its fertility.

Back to the land
An organic farmer seen with one of his most cherished raw materials – manure. Animal waste and plant compost should play an important part in the cycle of soil fertility. Organic farmers recognize this, and ensure that none of it is wasted. By contrast, on agribusiness farms, organic waste is a problem substance, something to be thrown away and replaced by artificial fertilizers.

people is no long-term solution and goes a long way to destroying the farming fabric of those hungry countries. Exporting our topsy-turvy way of farming has the same effect, as almost all aid agencies now recognize.

There is no sense in forcing the soil of North America, Europe and Australasia to produce more food, at the expense of the soil itself, if the high productivity cannot be kept up. By buying organic produce, and encouraging organic farming on a large scale, we will help not only ourselves but also the people who currently have no choice at all about what food to eat.

THE FACTS OF ANIMAL FARMING

There is nothing implicitly unhealthy in eating meat. Hunting societies like the Inuit or Eskimos have confounded nutritionists by surviving on a diet that is 100 percent meat. Yet in the same way, there are people, many millions of them in fact, who show the best of health while living on a purely vegetarian diet.

All this goes to show that humans are adaptable when it comes to food, and there is nothing biologically wrong with eating meat. What is wrong is eating meat that is produced from factory farms.

As with cereal farming, modern animal farming is geared to mass production but limited range. Gone are the goats, rabbits, ducks and geese – instead factory farmers concentrate on cattle, pigs and chickens. These are crammed together in "efficient" accommodation where a monotonous combination of food and climate ensures that they put on weight or lay eggs as fast as possible. The only countryside most of these animals see is through the ventilation holes of a truck on its way to the slaughterhouse.

Intensive beef and dairy production demonstrates how wholly unnatural animal farming has become. It is a well-established fact that calves should not be weaned for a week after birth because they need the colostrum present in their mothers' milk. But because the concern of today's farmer is the mother's milk production not bovine health, in practice calves are usually separated at birth and fed on milk powder and artificial feed after only a few days. All this, of course, takes place indoors. The resultant poor health of the animals means that their diet is heavily supplemented with antibiotics to counter the multiple infections to which they are prone. So, dosed with antibiotics, heavily sedated and often (in countries where it is still legal) injected with growth hormones, they will then be transferred to the "beef lot." Although these can vary enormously, they often have hundreds of animals ingeniously packed into crowded pens. The proportions of these boxes and the electric wires over their backs ensure that they do nothing but eat and defecate until they are ready to be slaughtered.

None of these animals ever has the experience of nibbling grass – in fact they never in their lives see a blade of grass. They are fed instead on silage and concentrates. They never have the security of any maternal affection, never know what it is to run and gambol with other calves, never experience grazing under a real sky while feeling the sun on their backs, or indeed the rain or cold of winter. They are never in their lives permitted to indulge any of the behavior patterns that are special to their species.

Pigs, if anything, fare worse. The pig, like most omnivorous animals, is highly intelligent. He has great curiosity, is capable of affection and anyone who has kept pigs in fairly natural surroundings will probably like and respect them. A factory pig is a sorry creature. Many suffer from poor circulation because the size of their hearts has been reduced by breeding (the heart is of little commercial value).

They then have to be treated with cardiac drugs to keep them alive, and tranquilizers to reduce stress – yet more chemicals to be passed on in food.

HOW FACTORY FARMING DAMAGES THE LAND

Eating factory-farmed meat is an inefficient way of nourishing yourself. A pig, for example, turns only one-quarter of what it is fed into pork. The rest goes into, among other things, keeping up its body temperature, maintaining its circulation and powering its muscles. By the same token, to feed an ox from calfhood to a finished beef animal takes from two to two and a half years and an enormous volume of food. In a factory farm, the great bulk of this is hay or silage, forage (green crops such as green maize cut by machine and carried to the beef lot), barley or other grain. Or, far worse, it can be high-protein food from tropical countries that need it far more than we do.

All of this animal food has to be extracted from the land (this takes energy; an animal does it for nothing) and it reduces the fertility of the soil it was

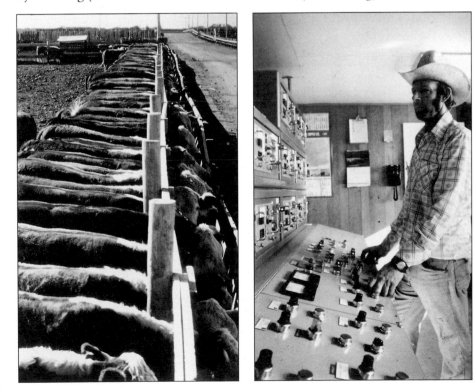

Push-button farming
Automation is the name of the game on the intensive animal farm. Here cattle are fattened up (*far left*) by bringing food to them in pens. What they are eating may well have been grown on the other side of the world. While they feed, the farm's activities are supervised from a computerized control panel (*left*). This is technology at its most inappropriate. Cattle will find their own food for nothing if allowed to do so. But they are kept off the land, and huge sums of money are spent on bringing food to them.

HOW PRODUCTIVE IS FACTORY FARMING?

How much more profitable is it to keep farm animals locked up? The answer in the case of hens is "not very." There are six basic ways of farming hens for eggs – these are shown below in order of intensity. With each system, the average egg output per bird per year is also shown. The more intensive the system, the more eggs each bird produces. But the increase is much smaller than you might expect: a battery hen,

for example, produces only about 35 "extra" eggs a year – just 15 percent more than a free-range bird. To get these extra eggs, the farmer has to make a huge investment in buildings, temperature-control equipment, other machinery and feed, and must also run a greater risk of disease. By contrast, free-range eggs cost very little to produce because the hens live largely on home-grown food.

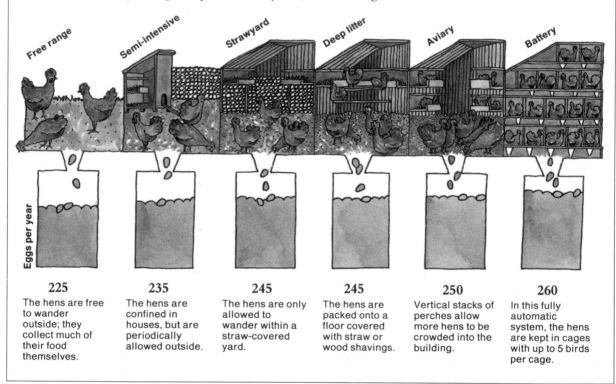

Eggs per year

Free range	Semi-intensive	Strawyard	Deep litter	Aviary	Battery
225	**235**	**245**	**245**	**250**	**260**
The hens are free to wander outside; they collect much of their food themselves.	The hens are confined in houses, but are periodically allowed outside.	The hens are only allowed to wander within a straw-covered yard.	The hens are packed onto a floor covered with straw or wood shavings.	Vertical stacks of perches allow more hens to be crowded into the building.	In this fully automatic system, the hens are kept in cages with up to 5 birds per cage.

grown on. The continual removal of all this vegetable matter inevitably impoverishes the soil until nothing at all will grow on it. The nutrients that are removed in this way are replaced in agribusiness by artificial fertilizer, at an enormous cost in fossil fuels. The land sputters back into life and can then go on growing more and more fodder to feed to animals.

The manure from all those animals shut up indoors is simply thrown away. In covered beef lots, it often drops down through slats on which the animals stand, to be washed along underground

channels with great expenditure of water and then into enormous lagoons. And there it stays, gradually percolating down through the ground to pollute waterways.

In open beef lots such as the ones that are seen in many American states, manure is pushed out of the pens with bulldozers and then piled up into enormous heaps – mountainous heaps. Eventually they disappear. Certainly they will never be used productively to enrich the land, because it costs too much to transport this material the hundreds of miles to areas where arable farmers would be able to use it. In theory anybody can come and help themselves to this bountiful supply of free manure but nobody does, partly because there are rarely growers in the proximity, partly because owing to some of the things that cattle are fed on the manure is considered unusable.

What has happened is that animals have been deliberately divorced from the land. The pattern of agribusiness animal farming is linear rather than cyclical. Artificial fertilizers produced by gas and oil wells are transported large distances before being applied to the land. Crops are grown with them which are then transported to animals held indoors. They are eaten and then voided by the animals, to end up in dead-end heaps or lagoons from which they are irrecoverable. The small proportion of these nutrients that does turn into food for humans ends up not back on the land but in the sea, as human sewage.

Wendell Berry, an American farmer and writer, has succinctly summarized the process. "Our agriculturalists have taken a solution and neatly divided it into two problems. The problem of fertilizing the land and the problem of disposing of the manure. If the animals were reared and fattened on the farms on which their food was grown, these two problems would not have arisen."

THE ALTERNATIVES TO FACTORY FARMING

The only possible justification that there can be for factory farming is that it is cheaper – a claim which is heard again and again whenever the inhumanity of the system is challenged. But it seems that agribusiness might have got its sums wrong.

Free-range pigs
The free-range pig (*right*) is a cheap and contented animal. All it asks for is dry shelter, some bedding, and a certain amount of extra food to supplement what it gathers for itself. Free-range pigs can be kept on a large scale with the type of housing seen here. In these conditions, they remain healthy and reproduce prolifically.

Factory-farmed pigs
When pigs are kept in crates (*left*), the cost of raising them leaps upwards. Although they may put on weight faster than free-range animals, they are more prone to disease. The set-up shown here is comparatively libertarian: many factory-farmed pigs never see the light of day.

POSITIVE ACTION
Everyday action on factory farming

- **Assume responsibility**
  Food producers take great care to ensure that factory-produced meat and eggs are never actually described as such. Shoppers therefore need to take on responsibility for identifying and avoiding factory-farm produce.

- **Look out for dishonest labels**
  If you do see factory-farmed food that is being sold under a completely deceptive label, complain about it. Many labels on factory-farmed foods are dishonest: battery eggs, to take a common example, should never be sold in boxes that show hens in the open.

- **Eat less meat**
  Excessive meat-eating is an easy habit to acquire. The conditions that animals are raised in would be greatly improved if average meat consumption fell – and more useful land would be freed as a result.

- **Support humane farmers**
  Because factory farming is carried out on such a large scale, its produce tends to be cheaper. Willingness to pay slightly more for food that is both free of chemicals and humanely produced encourages nonintensive farming.

In the most intensive factory farms, the costs of animal pens, feed transportation and constant medical attention are so great that the increased productivity is barely paid for. It's the same story as with biocides and fertilizers.

There is actually no clear cut-and-dried financial benefit brought about by locking animals up. Some types of factory farming may cost the same as keeping animals on the land, while some may work out very slightly cheaper. When factory farming is cheaper, the difference is often tiny. There is a marginal commercial advantage in keeping pigs very intensively. It is slightly more profitable to keep sows chained to the floor all their lives in cubicles so that they cannot turn round. It is slightly more profitable to take piglets away from their mothers as soon as they are born and rear them artificially. It is slightly more profitable to fatten pigs in total darkness so they cannot fight out of sheer boredom, turning the light on only when you feed them. But is it *worth* it?

In both Europe and America, a few farmers have decided that it is not, and that returning animals to conditions a little more like nature intended is actually good for them as well as their livestock. Others have gone much further, deciding that any form of factory farming is not only inhumane but dangerous. Factory-farmed meat is often full of antibiotics and other drugs. None of these entirely disappear before the meat is sold. By rejecting the use of "sub-therapeutic" drugs (ones used routinely without individual diagnosis) and by returning their animals to the land, these farmers produce safer meat.

This humane approach to farming may cost more in labor terms, but the cost of equipment is minimal, so it can work out cheaper. Pigs, for example, spend their entire lives out of doors or resting in small individual shelters and rearing their own young, while the porkers or baconers being fattened are kept in spacious, warm and comfortable strawed houses. Much of their food is free, and the animals do not suffer from diseases brought on by overcrowding and inactivity. Cattle also need less medical attention, while hens collect food that otherwise would have been unused. All of these animals improve the land enormously: they all produce valuable manure, while pigs throw in free plowing in return for their freedom.

If there were no other ways of producing meat and eggs then it might be argued that the gross and mass cruelty of factory farming is pardonable, although even then many of us would counter-argue that the ends still do not justify the means. But humane farming is viable. All it needs is support from the customer through the power of the purse. Everyone who eats meat is implicated in the crime of factory farming: if we all bought humanely produced meat whenever it was available, factory farms and their attendant cruelty would very quickly become a horror of the past.

4

FOOD FOR HEALTH—OR WEALTH?

Food without technology
Why simple and local is best
How to avoid eating additives
The wholefood alternative
Fast food and forests
Starting conservation in the kitchen

Like it or not, we live in the age of the super-market. For many people, the weekly visit to one of these caverns of commerce is an immovable feature of domestic life. The average shopper devotes the equivalent of over two work-ing weeks every year to pushing a wagon up and down the brilliantly illuminated aisles, and by the end of the year will have shoved several tons of purchases through the checkout.

Just what goes out of that wagon and into the car every week? Well, most of it is processed food – food that has undergone some sort of trans-formation before reaching the supermarket shelf. It may be bleached, colored, stabilized, flavored, vacuum-packed, heat-sealed, film-wrapped, canned, frozen, and, in some countries, even treated with radiation, but the one thing it is unlikely to be is wholesome. But why does so much of our food end up like this? Why can't we buy the real thing – without it being interfered with?

We might as well start with bread, because bread was probably the first food to fall into the hands of the processors, and the first food to be "rescued" from them by people who understood the bad effects processing was (and still is) having.

Western man has been living on bread for at least ten thousand years. For the first eight thousand, bread was made exclusively from the whole grain of wheat (or barley or rye) ground between two stones. But in about 150 B.C. Roman bakers started to tamper with the ingredients by using wheat flour that had been passed through cloth to sift out all but the fine white endosperm (the starch of the wheat grain). This latter contained most of the energy of the wheat grain, but none of the protein, none of the essential fatty acids, only one-fifth of the copper, hardly any of the manganese, zinc, magnesium and other trace elements that are essential to our health, or any of the B and E vitamins. These were all sifted out.

The appeal of this "new" bread, as indeed with all modern processed foods, was principally visual. Wealthy Romans liked it because it *looked* white and pure. After the fall of the Empire the Western world largely returned to wholemeal bread (in fact only a tiny minority of people had ever left it) and it wasn't until about A.D. 1000 that the rich people, and only the rich, began to use white flour again. Hair sieves, called temses, for sifting out the bran and germ of wheat, were introduced into England from

THE FOOD CHAIN
Part 1: *Where our food comes from*

Few foods escape some form of human interference before they reach the shops. Often the products we buy are unrecognizable, compared with their original ingredients, in taste, texture, flavor, and in nutritional value too. Undesirable changes occur at every stage of food's journey towards our stomachs.

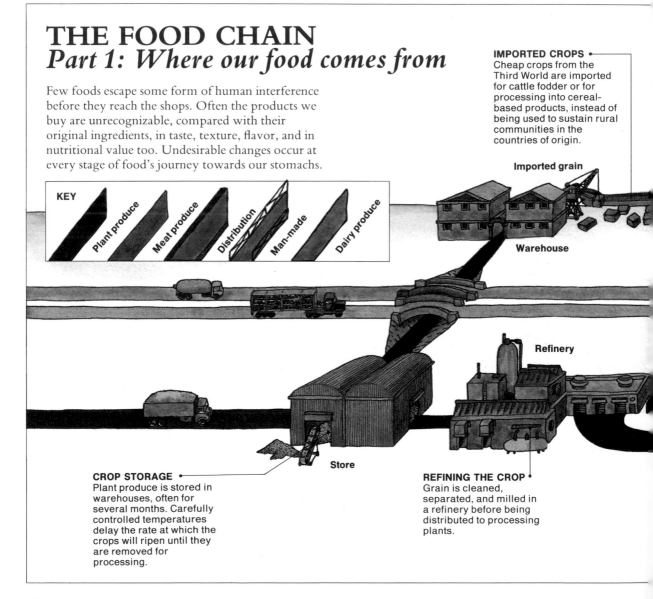

IMPORTED CROPS
Cheap crops from the Third World are imported for cattle fodder or for processing into cereal-based products, instead of being used to sustain rural communities in the countries of origin.

Imported grain

KEY
Plant produce
Meat produce
Distribution
Man-made
Dairy produce

Warehouse

Refinery

Store

CROP STORAGE
Plant produce is stored in warehouses, often for several months. Carefully controlled temperatures delay the rate at which the crops will ripen until they are removed for processing.

REFINING THE CROP
Grain is cleaned, separated, and milled in a refinery before being distributed to processing plants.

mainland Europe at about the time of the Norman conquest.

It was not, however, until about 1750, with the introduction of a device called the "bolter" into English wind and watermills, that the ball really started to roll. The wealthy and grand ate white bread, therefore everybody else wanted to eat it too. Bakers began to add alum, lime, chalk, and even the powdered bones of animals to try to make their loaves whiter-than-white.

The range for cheap and whiter-than-white bread culminated in 1960, when the Chorleywood baking process was invented. This turned breadmaking from a craft into a fully mechanized business. The result of this was, in many countries, to knock out nearly all of the craft bakers. In England, for example, it put baking in the hands of two huge companies, which turn out thousands of tons of wrapped sliced white bread a day. For a time, wherever factory bread became popular, the

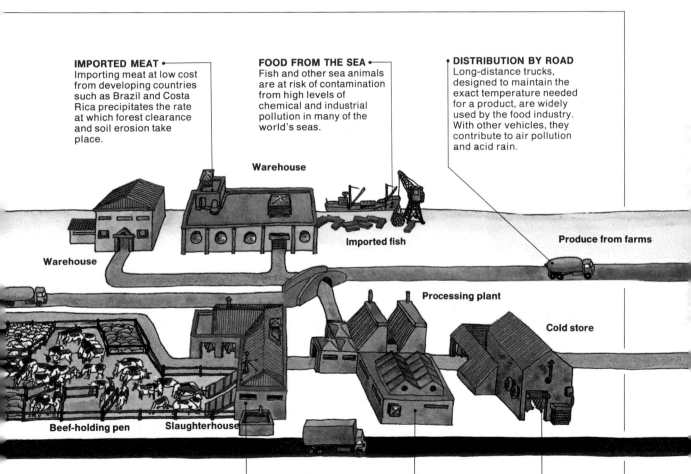

IMPORTED MEAT
Importing meat at low cost from developing countries such as Brazil and Costa Rica precipitates the rate at which forest clearance and soil erosion take place.

FOOD FROM THE SEA
Fish and other sea animals are at risk of contamination from high levels of chemical and industrial pollution in many of the world's seas.

DISTRIBUTION BY ROAD
Long-distance trucks, designed to maintain the exact temperature needed for a product, are widely used by the food industry. With other vehicles, they contribute to air pollution and acid rain.

Warehouse

Imported fish

Produce from farms

Warehouse

Processing plant

Cold store

Beef-holding pen

Slaughterhouse

SLAUGHTERHOUSE
Animals are often transported over large distances on the journey from the farm to the slaughterhouse, where they are either electrocuted or shot.

MEAT TREATMENT
Meat may go through a range of processes at this stage: bones and fat may be removed, low-grade meat, blood, and gristle may be added. What remains may then be cut up for storage.

MEAT STORAGE
Because of the high risk of bacterial infection, meat has to be frozen or carefully stored at cool temperatures.

non-processed variety lost favor and almost vanished without trace.

TECHNOLOGY OUT OF CONTROL

Food companies – the kind involved in wholesale processing – are fond of expressions like "convenience", "choice" and "meeting a public demand" when their activities need explaining. But the plain truth of the matter is that food processing is designed first and foremost to make money. The object of the Chorleywood baking process, for example, isn't to produce a better loaf: it's to cut the labor cost of baking, and to produce a very white loaf with large air holes, one that contains a lot of water and will keep for a long time. If a bread factory owner can sell a large loaf that is chiefly made from air and water he is not going to do badly.

White bread shows how processing robs a food of its proper nutritive value. As early as

THE FOOD CHAIN
Part 2: How our food is processed

Once food has been gathered and has undergone some basic treatments, it is transported to processing centers. There, with the help of the chemical and the packaging industries, it is transformed into the products we are used to seeing on the shelves in shops and in our homes.

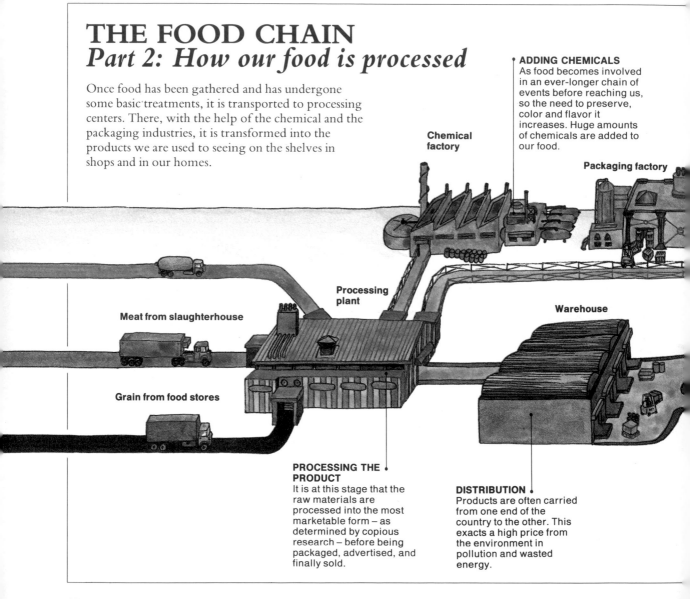

ADDING CHEMICALS
As food becomes involved in an ever-longer chain of events before reaching us, so the need to preserve, color and flavor it increases. Huge amounts of chemicals are added to our food.

Chemical factory

Packaging factory

Meat from slaughterhouse

Processing plant

Warehouse

Grain from food stores

PROCESSING THE PRODUCT
It is at this stage that the raw materials are processed into the most marketable form – as determined by copious research – before being packaged, advertised, and finally sold.

DISTRIBUTION
Products are often carried from one end of the country to the other. This exacts a high price from the environment in pollution and wasted energy.

1826 doctors were beginning to question the healthiness of a diet based on white bread. In that year a French physiologist named François Magendie reported the result of an experiment he had conducted with dogs, feeding some on pure white bread and water, and others on wholemeal bread and water. The animals fed on white bread died within fifty days; the wholemeal-fed ones lived on normally in quite adequate health.

In every single valuable constituent of diet, except calcium (in which white bread wins because of its added chalk) and chloride (which comes from added salt), original old-fashioned wholemeal bread is nutritiously richer than its processed counterpart. In addition to this, white bread is seriously deficient in fiber. Nutritionally, there is just no comparison between the two.

But good nutrition is not the point of processing food. The aim of the food processor is to take relatively inexpensive ingredients, combine

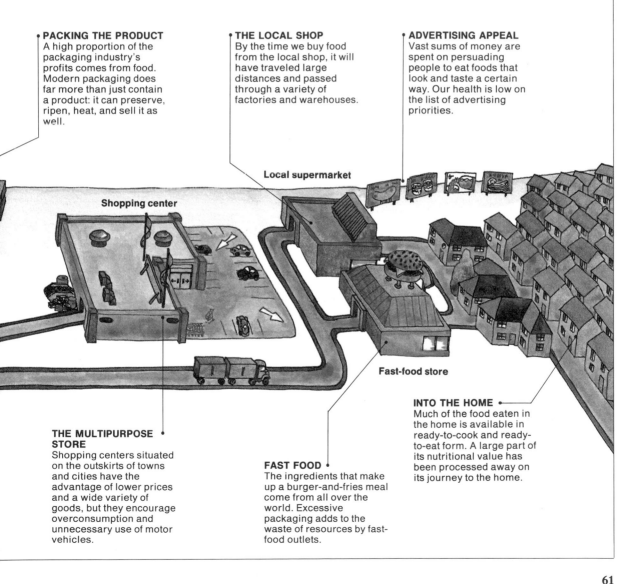

PACKING THE PRODUCT
A high proportion of the packaging industry's profits comes from food. Modern packaging does far more than just contain a product: it can preserve, ripen, heat, and sell it as well.

THE LOCAL SHOP
By the time we buy food from the local shop, it will have traveled large distances and passed through a variety of factories and warehouses.

ADVERTISING APPEAL
Vast sums of money are spent on persuading people to eat foods that look and taste a certain way. Our health is low on the list of advertising priorities.

Local supermarket

Shopping center

THE MULTIPURPOSE STORE
Shopping centers situated on the outskirts of towns and cities have the advantage of lower prices and a wide variety of goods, but they encourage overconsumption and unnecessary use of motor vehicles.

FAST FOOD
The ingredients that make up a burger-and-fries meal come from all over the world. Excessive packaging adds to the waste of resources by fast-food outlets.

Fast-food store

INTO THE HOME
Much of the food eaten in the home is available in ready-to-cook and ready-to-eat form. A large part of its nutritional value has been processed away on its journey to the home.

them, and then sell the result at a much higher price than the ingredients alone would have fetched. Once a formula proves successful, the "recipe" is used to produce an identical product year in and year out. This allows manufacturers to use brand names and from there the advertising and marketing machinery can set to work to create "brand loyalty," a crucial tactic in the battle for supremacy in the market place.

This leads to all sorts of garbage being foisted onto the consumer. There are nonfoods (foods which have little or no food value at all), fast foods (foods designed to be easy to transport and easy to serve up, rather than being beneficial to eat) and synthetic foods (foods which are made entirely from factory-created ingredients). Unpleasant though these may sound, skillful advertising ensures that we work our way through huge quantities of them, which encourages industrial agriculture, overfishing, overpackaging, and the manufacture of unsafe chemical ingredients.

MASS FOOD AND MASS WASTE

No one could pretend for a moment that anyone without a substantial and productive plot of land could do entirely without processed food. But it is important to realize that relying principally on processed foods is not only bad for your health: it is bad on a wider scale because of the ways in which processed food is produced.

Let us take bread again. There can be nothing simpler than gathering the seed of the wheat plant, grinding it, baking the resultant flour, and eating it. But processed bread is anything but simple. Its flour has to be of a standard type, and it is produced on agribusiness farms, with lavish use of oil, fertilizers, and biocides. Once the grain has been harvested, it has to be transported (again with great use of oil) and then ground into flour in huge mills where all manner of chemicals are added to it – chemicals produced by highly polluting chemical factories.

All processed food is the same: it involves the elaborate and costly waste of resources in order to produce something that is actually *inferior* to a straightforward product. It violates the principle of simplicity: the more complex the food and the

FROM NATURAL FOOD TO PROCESSED PRODUCT

Potato chips are an excellent example of the way in which the processing undergone by many foods reduces nutritional value but increases price.

RAW INGREDIENTS
Potato chips are made from specially developed varieties of potato which are high in starch and low in sugar.

PREPARATION
The skins are removed with water jets and caustic soda and the potatoes are then finely sliced. Light and air destroy nutrients during peeling and slicing.

COOKING
The potatoes are fried in vegetable oil. This is a more costly ingredient than the potatoes themselves. More than 30 percent of the weight of a bag of chips is the oil absorbed during frying.

ARTIFICIAL ADDITIVES
Artificial flavors, colors and antioxidizing agents are added to the chips after frying. They are sprinkled on to the slices by a revolving drum.

PACKAGING
The chips are dried, cooled and packed into bags. These are made from a variety of plastics. Most are non-biodegradable, and so contribute to the problem of packaging waste.

SELLING
The final product is 30 percent oil, together with substantial amounts of air. The price of the potatoes has increased by 500 percent through processing.

more ingredients that are listed in tiny letters on the side of the packet, then the more wasteful its production will have been.

The buying of factory-made foods like white bread also violates the principle of localism. In the manufacture of processed bread, economy of scale is very important for increasing profits. Both grinding mills and bread factories have therefore tended to get fewer and fewer down the years, and bigger and bigger, and consequently further and further apart. Bread has to be carried great distances nowadays, the manufacturing power is concentrated in a few hands, and local baking skills are gradually forgotten and lost.

For products where there is international competition for sales, the results of this mass production are quite ridiculous. At one time, every large town would have had its own beer. Now brewing is done by a few very big companies (mostly owned by huge food corporations). Every year they ship millions of gallons of beer (which is 99 percent water) from one country to another, while millions of other gallons travel in the opposite direction. The quality suffers because the beer has to be sterilized so that it keeps for months, and the extra fuel consumption and pollution is enormous. Where production has been local (as thank heavens some of it still is) all this nonsense is avoided.

This reckless use of natural resources is of no concern to the food corporations. They don't worry about overpackaging as long as it brings in the sales. The fact that they purvey food of little nutritional value doesn't worry them either; they can afford massive advertising to persuade people to eat and drink their products. Those other violators of the principles of simplicity and localism, the giant supermarket chains, love them, because they can flood their shelves with consistent products with a long shelf-life.

SIMPLE AND LOCAL IS BEST

Today's shopper is confronted by a baffling range of decisions and choices which our fore-bears never had to worry about. In the days before supermarkets, they simply bought what was in season, and that was that. There was

Spoiled for choice Today's shoppers have far more choices than they need. With such a huge range of goods for sale, choosing food can be a daunting and confusing experience.

no puzzling over different brands because often there was only one brand to buy.

But how can you make sensible choices when buying food nowadays without spending your whole life shopping? As an overall rule-of-thumb, when faced with a choice, buy the simpler and more local alternative. By doing this, you will select the product least interfered with by man-made ingredients, one that will be best for your health and that of the world beyond your kitchen. Furthermore, you will be buying the product that has used least fuel (and therefore caused least pollution) on its journey to the shop.

THE ARTIFICIAL INGREDIENTS IN PROCESSED FOOD

Food manufacturers have alchemic powers: they turn base metal into gold. Or rather, with the help of a little chemistry, they turn a perishable short-lived raw material into something longer-lasting and easier to sell.

There was a time when the only food additive was salt, but that time is long past. Now there are hundreds of synthetic substances that can quite legally be put into food. There are the preservatives, which are used to stop food from rotting on its long journey to the shops, and then there is a host of other chemicals which can probably be best described as food cosmetics: they "make up" food to look and taste better than it really is.

HOW PROCESSING DEVALUES OUR FOOD

In an unprocessed state, most foods are excellent sources of protein, fiber, vitamins, and minerals. Furthermore, they are usually low in polyunsaturated fat and salt, and do not contain refined sugar. Yet increasingly we choose to eat food that has been processed in some way. Artificial colors and flavors, antioxidizing agents, preservatives, and stabilizers are frequently added. Many of these are potentially hazardous. This table shows how some common foods, which are easy to buy and eat in their unprocessed form, are usually bought and eaten in a highly processed form instead.

MEAT
Unprocessed meat is a good source of protein, vitamins and minerals. Grilled, roasted, or stewed lean meat is low in saturated fat.

MEAT PRODUCTS
Products such as sausages, pies, hamburgers and pâtés are high in saturated fat and low in other nutrients.

FRUIT
Fresh fruit is rich in Vitamin C, fiber and minerals. It can provide an immense range of color, flavor and texture.

FRUIT PRODUCTS
Artificial additives often replace lost flavor and color. The fibre and vitamin content is depleted.

FISH
Grilled, baked, or poached fish is an excellent source of minerals, vitamins and polyunsaturated fat.

FISH PRODUCTS
Artificial additives are often used to replace lost flavor and color. Water, bones and skin are added to increase bulk.

VEGETABLES
Potatoes, to take just one example, are an excellent source of protein, fiber, low-fat carbohydrate and vitamins.

VEGETABLE PRODUCTS
When processed, nutrients concentrated just below the skin are mostly destroyed. Most potato products contain artificial color and flavor.

MILK
Fresh whole or skimmed milk is a rich source of calcium, protein and Vitamins B_2 and B_{12}.

MILK PRODUCTS
Ice cream, milk-shakes, and cheese spreads are high in fat, sugar, and salt. They often contain artificial color and flavor.

Like our close relatives, the apes, we select food mainly by sight: color is very important to us. This has not been lost on the food technologists. They have developed a huge array of artificial colors that can be tuned as the controls of a television set. With them they brighten, deepen, or lighten the color of a food to make it as appealing as possible. As a result, people have now forgotten what real food looks like. They have also forgotten that food that has been stored will *never* look as bright as food that is fresh – unless someone has meddled with it.

Having "engineered" the color, the manufacturers then use artificial flavorings to replace the flavor that is lost in processing. They liven it up again splendidly. They can also mimic the flavor of a natural ingredient which can then be left out altogether – much cheaper than using perishable, unreliable, real ingredients. If you look at the label of any fruit-based processed dessert you will probably find that the cherries, strawberries, apples, or whatever it is you are looking forward to are simply not there. And just because a can of soup says "chicken flavor" on it, that doesn't mean to say that it will actually have any chicken in it.

Finally come the texture controllers – emulsifiers, thickeners, gelling agents, extenders and stabilizers, to name but a few. These create the most marketable texture by preventing lumps or crystallization, either too much or too little moisture. They ensure that the elements put together by the food technologists stay blended and do not curdle or separate.

With all these at their disposal, food companies turn something cheap and unpalatable into a substance that the shoppers (after being subjected to the right advertising) will clamor for. They can, for example, take plant oils and milk by-products, bubble hydrogen through them, add artificial colors, flavors, and emulsifiers, squirt the resulting mass into plastic tubs or wrap it in brightly colored paper, and sell it as margarine, an "improvement" on butter. And because margarine is produced in industrial plants on a huge scale, it can be manufactured very cheaply, and consequently it finds an enormous market.

Food marketing and advertising is one of the manufacturer's most important investments. What

---

POSITIVE ACTION
Avoiding artificial additives

- **Read the label**
  Looking at the labels on cans and packages will make you aware of which products are most heavily processed.

- **Buy fresh produce**
  Food that you buy fresh and cook in your own home will contain only those additives you intend to be there.

- **Know the risks**
  Learn which additives are known or believed to cause health problems and avoid buying products which contain them.

---

exactly is in the food being sold is of little relevance to the advertising agents. They are much more concerned with catchy jingles to ensure a brand name sticks in our minds and glossy packaging that promises satisfaction within. The actual ingredients in a product are the smallest item to feature on the label. And why is that? For the understandable reason that manufacturers do not want to admit what has gone into it.

ARE ADDITIVES *REALLY* DANGEROUS?

The use of artificial additives in food has increased tenfold in the last thirty years. In the United States alone it is estimated that we now eat more than 6 pounds every year. Now anything that you eat in that sort of quantity must have *some* effect on the body. Exactly what that effect is is still not certain.

Adults at least have some powers of discernment when it comes to choosing food, despite the influence of advertising. But children have few, if any. Encouraged by the advertising man, they consume sweets, chocolates, potato chips, and all manner of processed food in great quantities. It is precisely these products that are most heavily burdened with additives.

It is now thought quite probable that intolerance to these chemicals might be a significant factor in illnesses that commonly affect today's children, illnesses such as hyperactivity and skin disorders.

Doctors in America and in Britain have carried out experiments to ascertain whether or not additives are to blame by feeding children on additive-free diets. In many cases, the problems disappeared. An American clinician specializing in allergies conducted a range of experiments in the 1970s and concluded that half of all hyperactive children could be cured simply by eliminating certain chemical additives from their diet. Some additives have been withdrawn, but in many cases children in one country are eating additives which are banned in another.

With so many additives in food, it is difficult for anyone to avoid them entirely. Adults too accumulate these toxic substances in the body. Nitrites are just one example. They are used to preserve meat, and are a possible source of cancer. Their use is now restricted in many European countries, but although nitrite-free meat products are available in some places, approximately 5 million tons of cured meat and fish are still being treated with nitrite in the United States alone.

Americans have gradually reduced their consumption of nitrites, and as they have done so the incidence of stomach cancer has also fallen. In Japan on the other hand, where nitrite is consumed in large quantities, there is the highest incidence of stomach cancer in the world. No doubt at some time in the past the "experts" would have described nitrites as perfectly safe.

PRESERVED FOODS AND HOW TO CHOOSE THEM

It has been said that the only good food is food that goes bad. There is a lot of truth in this. Food is organic: in an unadulterated state it will decay naturally, and the only possible justification for interfering with food is to slow this process down.

Food preservation is an ancient art, one which until quite recently relied on natural substances like salt, sugar, spices, vinegar, and the drying power of the wind. Most households would dry or pickle a variety of foods in autumn, and this would make up for the lack of fresh food until the following spring. But once spring was well under way, no one would dream of eating preserved food in preference to fresh food now that it was available.

HOW FOOD IS PRESERVED

Food preservation nowadays is so widespread and efficient that we take the year-round availability of most foods for granted. This table shows the six main ways of preserving food and explains how they work and what harmful effects they may have on our health.

METHOD	EFFECT
CANNING	Food is sealed in airtight metal containers. The can and its contents are then heated so that the food is cooked and sterilized. The nutritional value is greatly reduced. Artificial additives often replace lost flavor and color.
FREEZING	Very low temperatures arrest the natural process of decay. Frozen food often has better levels of vitamins than food sold as fresh, because freezing prevents the natural decrease after harvesting.
DRYING	Moisture is removed by applying heat and air. This is either carried out naturally, by the sun or wind, or mechanically. Artificial additives are used in large amounts to enhance processed dried foods.
BOTTLING	Food is sealed in airtight glass containers and then sterilized by heating. In some cases this process can improve the natural flavor. Bottling involves minimal hazards to health.
IRRADIATING	Large doses of ionizing radiation preserve food by delaying the maturation process and by killing insect pests and micro-organisms. Irradiation also produces chemical reactions in food which are suspected of causing cancer.
VACUUM-PACKING	Food is put in plastic bags and the surrounding air is then sucked out, creating a bacteria-free vacuum around the food. The plasticizers in many of the wrappings can cause cancer. The risk is greatest with fatty foods.

Nowadays, we eat preserved food all year round. Some of it is still prepared with old-fashioned methods such as salting and pickling and bottling, but more often food is preserved either by lacing it with chemical preservatives or treating it in some way – canning, freezing, or even treating it with radiation. But if you want to avoid food that does contain chemical preservatives, which of the treatment methods is safest?

Canning almost qualifies as an old-fashioned method of food preservation – it was invented in the nineteenth century and has been in constant use for decades. Canned food is heated after the can is sealed to kill the bacteria that cause decay. Although the heat treatment also destroys substances such as vitamins, it is, on the whole, fairly safe. However, there are two reasons to avoid overconsumption of canned food. First, cans are an appalling waste of metal. Hardly any of them are recycled, although many of them could be (see p. 89). Second, they require the use of a solder to keep them airtight, and this often contains lead. Anything that encourages the use of poisonous metals cannot be good in the long run.

Provided it is not kept for an excessively long time, frozen food is a much better alternative to chemically preserved food. The only major drawback of frozen food is the amount of energy it uses in storage – in a large supermarket (especially where the freezer cabinets are open-topped) vast amounts of electricity are needed to keep frozen food frozen.

Irradiated food (which is banned in some countries) has been represented as a great step forward in food preservation. But its long-term effects, like those of food additives, are quite unknown. It may succeed in arresting the process of bacterial decay, but no-one is really sure if the ensuing chemical changes in the food are entirely safe. Irradiated food is only for those who wish to conduct experiments with their health.

ARE ADDITIVE TESTS ADEQUATE?

It has been said that one of the most reliable ways of testing the safety of food additives is to toss a coin. This is because animal tests – the kind used to detect carcinogenic (cancer-inducing) chemicals – are accurate in less than 40 percent of total cases.

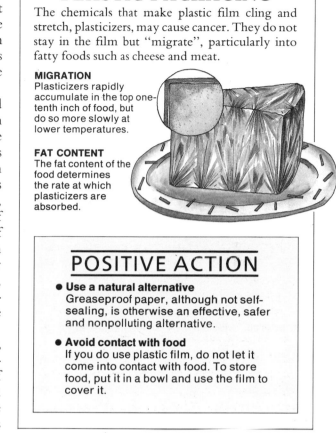

HEALTH HAZARDS FROM PLASTIC PACKAGING

The chemicals that make plastic film cling and stretch, plasticizers, may cause cancer. They do not stay in the film but "migrate", particularly into fatty foods such as cheese and meat.

MIGRATION
Plasticizers rapidly accumulate in the top one-tenth inch of food, but do so more slowly at lower temperatures.

FAT CONTENT
The fat content of the food determines the rate at which plasticizers are absorbed.

POSITIVE ACTION

- **Use a natural alternative**
  Greaseproof paper, although not self-sealing, is otherwise an effective, safer and nonpolluting alternative.

- **Avoid contact with food**
  If you do use plastic film, do not let it come into contact with food. To store food, put it in a bowl and use the film to cover it.

The task of testing an additive for safety is complicated, time-consuming, and costly. To save money, the tests are designed to reveal only certain illnesses such as cancer and diabetes. Many reactions like asthma and migraine go unmonitored, so it is just hard luck if you find yourself suffering from them – the blame is seldom placed where it is due.

As with biocide assessment (see pp. 44–6), the tests are all short-term. However, cancer can take from twenty to thirty years to emerge, so this makes the current methods of testing ridiculous. Indeed, it is quite common for additives to have been in use for several years before the related health risks, such as arthritis, depression, eczema, or acne come to light.

It is practically impossible to gauge the levels

THE UNHEALTHY KITCHEN CUPBOARD

Many of the foods with which we keep our kitchen cupboards so well stocked are not as nutritious as the labels might suggest. Processed foods are often full of chemicals: these may color, flavor, preserve, and stabilize very well, but they have little nutritional value and in some cases may be harmful to health.

SOFT DRINKS
Carbonated beverages are basically sugary water solutions with artificial color and flavor added. Many of these additives cause allergic reactions and hyperactivity in children.

CONDIMENTS AND PRESERVES
Most commercial pickles, spreads, jams, and ketchups contain large amounts of sugar, artificial color and flavor, emulsifiers and preservatives.

PROCESSED FRUIT AND VEGETABLES
Canned fruit and vegetables may be colored artificially with one of 14 azo dyes which can cause allergic reactions, particularly in children. Dried produce is often preserved with sulfites and sulfur dioxide. These too can cause allergic reactions. Dried vegetables may also include large numbers of chemical additives which make them appear "natural" when reconstituted.

CANDY AND DESSERTS
Confectionery and packaged desserts are laden with sugar, fat, artificial color and flavor, antioxidants, and preservatives. Some of these are suspected of causing cancer and allergic reactions. They are also very low in vitamins, minerals, fiber, and protein.

CEREALS AND BREAD
White bread and cereals made from refined flour usually contain bleaching agents, preservatives, and antioxidants which in some cases cause allergic reactions. Nutritionally they are greatly inferior to those made from wholegrains; in particular, the valuable fiber content has been removed by processing.

POSITIVE ACTION

How to make your cupboard healthier

- **Buy fresh produce**
  Fresh fruit and vegetables are likely to be the least tampered with. Otherwise use frozen products or those canned without sugar, salt, or chemical additives.

- **Choose wholegrains**
  Cereals, bread, biscuits, pasta, and rice made from wholegrains are rich sources of fiber, carbohydrate, B vitamins and protein.

- **Eat flesh not fat**
  Choose lean cuts of beef, poultry, or fish, all of which are low in fat and high in protein. Fish is a valuable source of polyunsaturated fats, vitamins, and minerals.

- **Choose natural drinks**
  Fruit and vegetable juices are usually free from added sugar, flavorings, or colorings. Organic wines and ales, which are grown and fermented without chemicals, are now widely available.

- **Avoid excessive sugar**
  Choose natural yogurt, dried or fresh fruits, seed- and nut-based snack bars, wholemeal cakes and biscuits made with unrefined sugar or honey.

- **Avoid excessive salt and flavorings**
  Choose from the wide range of pickles, jams, ketchups, and sauces that are free from excessive sugar, salt, and without artificial color, flavor, and preservatives.

MEAT PRODUCTS
Most meat is high in saturated fat, which has been linked with heart disease. Meat products also contain large amounts of salt and artificial additives. Some of the preservatives used in meat products are suspected of causing cancer, asthma, and reproductive problems.

ALCOHOLIC DRINKS
Large-scale brewing and wine-making involve the use of chemical additives, some of which have caused or are suspected of causing health problems. Sulphites, for example, used to preserve beer and wine, may cause allergic reactions and reproductive difficulties.

at which additive intake is harmless, because the cumulative effects of additives consumed over a number of years are as yet unknown. Furthermore, many substances began to be poured into the factory food mixers long before additive legislation appeared, and were therefore never even tested in the first place. The risks for these are quite unknown.

To make matters worse, there is one threat which is even less understood, and that is the "cocktail effect." This is what happens when you eat any processed food that contains more than one artificial ingredient. The additives are all mixed together in the body and, from then on, it is anyone's guess how they will interact with each other. Additives are only tested individually. If you are trying to market a food additive, you are only required to supply confirmation that your product is safe *on its own*. In practice it will rarely be eaten except in combinations – sometimes with as many as twenty different substances. The effect of these chemical cocktails is again unknown, but common sense suggests that if anything, they will be more, not less potent.

HOW SAFE ARE UNPROCESSED FOODS?

You can see and smell food that is rotten, but you cannot see or smell food that is contaminated with biocides. In the summer of 1985 six Californians died after eating locally grown watermelons. Altogether 1,350 people fell ill from the same cause. Hours after eating what looked like perfectly wholesome fruit, they suffered seizures and loss of consciousness. Investigators claimed that aldicarb sulfoxide, manufactured by Union Carbide under the brandname Temik, was responsible. It had been sprayed on to the melons to protect them from pests. So the cause of illness was biocide poisoning.

Little good comes of cutting down on processed foods if the fresh alternatives are unsafe. The California poisoning shows that unprocessed foods can also be dangerous to our health. Major biocide accidents such as this may still be rare, but biocides in foodstuffs can affect many of us at subclinical levels. In other words, not severely enough to provoke symptoms in the short term but sufficiently to cause long-term health problems.

Biocide residues are routinely found in a great variety of fruit and vegetables. Biocides find their way into foodstuffs in various ways. Herbicides and fungicides are used on most commercial farms, and these infiltrate into crops through the plants' roots. Fields of wheat are sprayed up to eight times with herbicides and fungicides during the growing season. More biocides are then applied during storage. Up to 40 percent of the poisons in the grain are likely to survive milling and baking and end up in a loaf of bread. Even wholemeal bread may contain biocides used on the farm.

Biocides are not the only problem. Excessive use of nitrates is leading to a potentially harmful build-up of residues in vegetables. A recent survey in Britain indicated that 30 percent of the fruit and vegetables on sale had detectable residues. In Germany lettuces have been shown to contain as much nitrate as 8 gallons of drinking water.

The lavish use of nitrates also greatly increases the water content which the consumer will eventually have to pay for in the shop. Apples, for example, contain greater quantities of water if they have been grown in orchards that have been dosed with large amounts of artificial fertilizers. "Golden Delicious," a variety widely grown in France and Italy for the European market, respond particularly well to such fertilizer applications. The apples can in effect be pumped full of water as they mature, thus bulking up the size and giving them the shiny appearance that is so attractive on the supermarket shelf. They may taste horrible but that doesn't seem to matter. Much of the vegetables and fruit that look so fresh and healthy on the shelf are in reality heavily contaminated. Indeed, they look the way they do often *because* of the chemicals that have been sprayed on them.

THE WHOLEFOOD ALTERNATIVE

There is nothing mysterious about wholefood: it is simply food that does not contain additives, has not had most of its raw ingredients removed, and is not sold under a cloak of deception. Wholefoods are simple, unrefined, and, where possible, they are locally produced.

The wholefood movement is concerned with restoring the idea that what goes into food is more important than the way it is dressed up. Because wholefoods are variable – they don't always come

in standard shapes, sizes, and colors – and because they are not produced by the large-scale suppliers, supermarkets have tended to ignore them. But the power of the purse is such that they are now admitting wholefoods on to their shelves as fast as they possibly can.

On a purely personal level, wholefoods are the healthy alternative. The "diseases of affluence" that people in the West suffer from – heart disease, diabetes, stomach and bowel cancer – are closely linked to our eating habits. The new consensus among nutritionists is that protein, once the dieticians' favorite, has been wrongly overvalued, and that unrefined carbohydrates and roughage (the principal ingredients of wholefoods) are much more valuable. So we need a shift from the high-meat, high-fat, high-sugar-and-salt, low-fiber diet to one a little more like that of days gone by when lack of refining and processing was the keynote.

THE VALUE OF UNREFINED FOOD

It is one of the marvelous things about nature that the edible stuff it produces is both full of nutritional value as well as being good to eat. If the diet is composed of unrefined "natural" food then it is not even necessary to have a particularly wide variety of foods. The Borana tribe in northern Kenya live almost entirely on milk – with a little meat from time to time – and these people are very well and very healthy. Why don't they suffer from thrombosis and heart disease as we are told we will if we eat too much fatty dairy produce?

Well, the first reason is that their staple food – milk – comes straight from the cow, and their cows do not begin life by being snatched from their mothers and being fed on such things as cattle cake and antibiotics. Furthermore, the milk they produce is not processed in any way. It is not taken off in tankers and transported here and there before being dumped into great metal vats and whisked up by metal paddles. It is not treated with heat to kill off bacteria or packed into cartons and then stacked up on supermarket shelves.

Second, the Borana don't suffer from coronary thrombosis because their lives are physically active. Similarly, the men who built the great Gothic cathedrals or the vast stretches of railway track got their strength from unrefined foods.

This isn't a plea for everyone to drink milk. There are of course many people, like the Chinese for example, who eat scarcely any dairy products at all. There are also people whose diets rely heavily on some other food. But their food is nevertheless *unrefined*. Food that is as close to its natural state as possible is, generally speaking, not going to do you anything but good.

In the Western world we have been slow to learn the simple lesson that wholefood is better than food which has had half its goodness refined away. Our ancestors lived on a diet of grain, fruit, and nuts, but it was not until the beginning of this century that the beneficial effects of this diet were rediscovered by Dr. Bircher-Benner in Switzerland and given the name of muesli. Humans did not evolve on a diet of canned foods and processed meals: doing without them is the common-sense way to stay healthy.

WHOLEFOODS AND THE LAND

There is another benefit to be gained by eating wholefoods – one which extends back along the human food-chain from consumer to producer and then to the land. Where possible, wholefoods are made from the produce of organic farms (see pp. 42–3). This means that eating wholefoods does not encourage the use of farm chemicals, or the overproduction of crops at the expense of both land and wildlife. Because it cannot be stored for long periods, most organic produce has to be local in origin, so eating wholefoods also helps to reduce the waste involved in transporting food around the world to places that should already be self-sufficient.

Many people who eat wholefoods are vegetarians, but this certainly shouldn't prevent any meat-eater from insisting on wholefoods instead of their processed equivalents. Quite the reverse. The more people who eat wholefoods the better, the more land will be farmed organically, the less biocides will be used, and the less the world will be cluttered up with truckloads of factory foods speeding toward the supermarket shelf.

THE FAST-FOOD PHENOMENON

Like cars and record players, hamburgers – the great standardized food – are assembled from parts that come from all over the globe: they are about the least "local" food that you can possibly buy. If you eat a hamburger in England, for example, the meat may well come from English cattle fed on American maize and soya from Brazil, or even from Thailand. The bun will be baked in England, but the wheat in it will come from North America. The cheese (processed) will be from Holland, the onions from Spain, the tomatoes from Italy, and the lettuce from Spain or California. Although the french fries will probably be made from local potatoes, the wrapping around them will be made

A fiber-rich diet
Eating foods rich in fiber does more than satisfy the appetite, it also reduces the incidence of many digestive diseases such as cancer of the colon. Fiber-rich foods have environmental advantages as well. Because they do not undergo extensive refining, they require less energy to produce than processed foods. They also generate less waste.

from Scandinavian paper. The recipe, of course, is American.

There are now nearly 40,000 hamburger outlets in the world, 9,000 of which belong to McDonald's. McDonald's proudly describes their food as being "adjusted to technology," and they achieve this technical perfection by an almost military attention to the details of size, weight, ingredients, packing, and presentation. A McDonald's manual ensures that a McDonald's meal is the same the world over. Even the sales attendants are standardized – packaged in McDonald's paper hats, nylon shirts and trousers.

The operation is hugely profitable. If you had had McDonald's shares in 1966, you would have seen their value grow fifty times over in seven years. In 1984 McDonald's had sales of over $10 billion, and by 1985 they had sold over 50 billion hamburgers. If McDonald's lined up all the hamburgers sold between 1955 and 1984 they would circle the equator over one hundred times.

But why do we eat all these hamburgers? Why have even the French with their fastidious culinary habits taken to them, with dozens of hamburger shops in the Champs-Elysées? What makes them so compelling that Austrians, Kenyans, Filipinos, Italians, Taiwanese, and even Japanese have all become McDonald's fans?

The main reasons are that the taste is engineered with the greatest possible precision to appeal to everybody, and the whole operation is geared to speed. It has proved to be an enormously successful product. But since it provides us with cheap and instant food, what, you may ask, is wrong with that?

THE GROWTH OF FAST FOOD

Fast-food stores have sprung up all over the world since the fifties, and hamburgers in particular have gained immense popularity. Beef consumption has risen proportionately. This table shows how rapidly sales of hamburgers from one leading chain have risen over a period of 20 years. For the countries that produce the meat for fast food, this has had catastrophic results. Tropical forests are being cleared at an alarming rate to make space for grazing land for cattle. Although over half of Central America's arable land now produces beef, the amount of it eaten by local people is actually falling every year.

YEAR   **HAMBURGERS (billions)**

1965
Forest clearance began to make room for cattle ranches. It takes 15–20 pounds of grain to produce one pound of beef.

3

1975
By now, sales of hamburgers were doubling every five years. The number of beef lots and cattle ranches in South America rose accordingly.

18

1985
Each second 140 hamburgers are bought from just one of the world's fast-food chains. Sales are increasing by 10 billion every year.

60

FAST FOOD AND FORESTS

Because hamburger corporations are multinational, they can shop around for the "components" that go into their products. The component of a hamburger that is most important, in terms of how it affects the environment, is meat. In the United States, to take the largest consumer first, a considerable proportion of the hamburger meat originates on beef ranches in Central America. In Costa Rica, for example, 42,000 tons of beef are exported each year, mainly in order to supply to the many firms like McDonald's.

The tragedy is that Costa Rica is completely unsuited for cattle farming. The land on which the cattle must graze has to be created by clearing vast areas of forest, and between 125,000 and 175,000 acres of forest are destroyed annually to provide short-term pasture. The removal of the forest cover leads to soil erosion and much of the new pasture turns to scrub within a few years. As a result, one-third of the country is now covered with pasture or infertile scrub. Small farmers are squeezed out as large ranches expand, but very few jobs are created on beef ranches because it takes only one person to look after a thousand cattle. This has perilous consequences for the local economy.

Although fast-food chains won't own up to using beef from Central and South America (aware that this is very unpopular with the United States ranching and feedlot interests), there is no doubt that this "hamburger connection" is a reality. Costa Rican ranchers have readily confirmed it.

But European hamburger eaters have no cause for complacency. Although European beef comes from Europe, the cattle are fattened on soya meal, half of which comes from Brazil and other Third World countries. Now a very high proportion of Brazilian soya – nearly nine-tenths of it – is grown on land formerly farmed as small holdings by peasants. These small parcels of land were once productively worked with intensive labor, and provided a means of subsistence for many people. The new soya farms are worked by huge machines, throwing people out of work, and also consuming more tracts of forest as the demand for soya rises.

Fertilizing the soil
A Brazilian farmer (*above*) tries to revitalize ailing plowed forest soil with artificial fertilizer. This is the beginning of a spiral in cost and production which will eventually exhaust the land.

The pace of forest clearance
Vast areas of tropical forest in Brazil (*left*) are cleared to create land for cattle grazing. The exposed soil quickly becomes seriously debilitated and is prone to erosion.

POSITIVE ACTION

Helping to localize the food industry

- **Buy locally**
  Wherever possible buy food that has been produced close to where you live. This saves on unnecessary storage, transportation, and distribution and therefore saves on energy, fuel, and overheads.
- **Ask questions**
  Find out where the meat you buy has come from and what methods of farming were used. Support farmers who use humane methods and supply locally.
- **Read labels**
  Notice where the food you are buying has come from whenever possible. Avoid produce that has been imported from abroad when home brands are also available.
- **Encourage change**
  Ask food retailers to provide food that has been produced locally, or at least without causing environmental damage in the country of origin.

WHAT HAPPENS WHEN A RAIN FOREST IS CLEARED?

As the demand for beef in the United States, Europe, and Japan goes up, the pressure on the countries of Central and South America to produce more beef increases. Countries like Honduras, El Salvador, Guatemala, Panama, and Nicaragua have all established pastures by logging or burning their forests. By 1985 two-thirds of Central America's accessible rain forests had been cleared or seriously depleted, all to feed cattle for wealthy westerners.

Tropical rain forests are being cleared at a rate of 7,700 square miles every year, not to produce permanent farms, but to make a short-term profit before the land is ruined. This is an act of extreme foolishness because tropical forests contain thousands of species of plants and animals which are of critical importance for a wide range of scientific research (see p. 111). Panama, a tiny country, has as many plant species as the whole of Europe put together. As the forests are annihilated so are these invaluable forms of plant life, and so too are the forest Indians who are the world's real experts on the jungle environment.

The soil in these countries is useless without the tropical forests. Most of the nutrients are not in the soil, but above ground, locked up in the trees and plants. When anything dies, it is not absorbed into the soil but rapidly converted into new plant growth. There simply isn't any fertile humus – the stuff that you can scoop up in great handfuls in any temperate woodland area. And once the multistory canopy of the trees is removed, the soil is exposed to torrential rainfall and the fierce heat of the sun, and erosion quickly sweeps it away.

There are no such things as "old established" farms on this so-called reclaimed land. They don't last that long. Between 1966 and 1983, 25 million acres of Amazonian forest were converted to pasture. Yet by 1986 nearly all the ranches that were cleared before 1978 had been abandoned because soil erosion and loss of productivity made them worthless.

STARTING CONSERVATION IN THE KITCHEN

It's not only fast food that is responsible for this habitat damage: there are plenty of products such as canned meat, fruit and imported animal feed which also play a part in this devastation. In Thailand, millions of acres of tropical forest have been cleared to produce meal for European pigs. In Niger and other West African countries where deforestation is almost total, peanuts were exported during the last Sahelian drought to keep French beef and dairy cattle well fed, even when hundreds of thousands of people were starving.

Tropical forest conservation begins in the kitchen. There is no point in deploring the fate of the forests, then digging into a beef sandwich produced by Brazilian cattle feed.

Forest destruction takes place because the food industry has turned localism on its head. It imports food from far-flung places where the local population has none to spare. To put a stop to this, all that is needed is a little persistence. There is no reason for anyone to give up fast food, but they should make sure that what they are eating is thoroughly home-grown. By finding out where food comes from, and then rejecting that produced in endangered environments, this short-sighted trade can be brought to a halt.

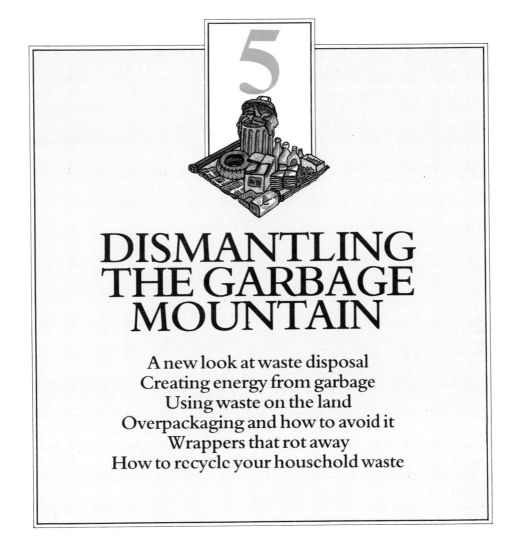

DISMANTLING THE GARBAGE MOUNTAIN

**A new look at waste disposal
Creating energy from garbage
Using waste on the land
Overpackaging and how to avoid it
Wrappers that rot away
How to recycle your household waste**

Anyone who has ever excavated an old country garbage dump – in order to turn the site into a rose garden perhaps, or to build a shed – will probably have been interested in what was in it. All you will usually find in such dumps are heel irons (those horse-shoe shaped irons that farming men nailed to their boots), the occasional decayed old boot itself, old enameled saucepans with holes in them, bits of pottery, the almost completely worn-out remnants of sickles, or scythes, or spade blades, and, in the upper horizons of the heap, maybe some broken bottles.

And that will be it: decades of garbage in just one small pile. Now there were no trash collections in those days – people had to dispose of their own

garbage, so these dumps contained everything that was thrown away. But all organic material, except leather, went to the pig, or if there was no pig, to the compost heap. Of real garbage, of the kind we produce today, there hardly *was* any!

By contrast, we live in an age of unprecedented waste. We generate garbage at the rate of between 45 and 165 pounds a month. Of course, we produce more garbage today than ever before in the history of mankind.

When the people of the future dig down into *our* garbage dumps, they will get a fascinating and voluminous insight into our consumption habits. Although the paper, cardboard, and kitchen leftovers which make up over half of our household

THE WORLD OF WASTE
What happens to our daily garbage?

As we continue to produce garbage in ever-increasing quantities, the ground, the sea and even the air around us are becoming contaminated by it. Most household and industrial rubbish is an unwholesome mixture of different materials, from harmless organic matter to highly polluting plastics and poisons. Little use can be made of waste in this form, and as a result the good part of waste gets thrown out with the bad.

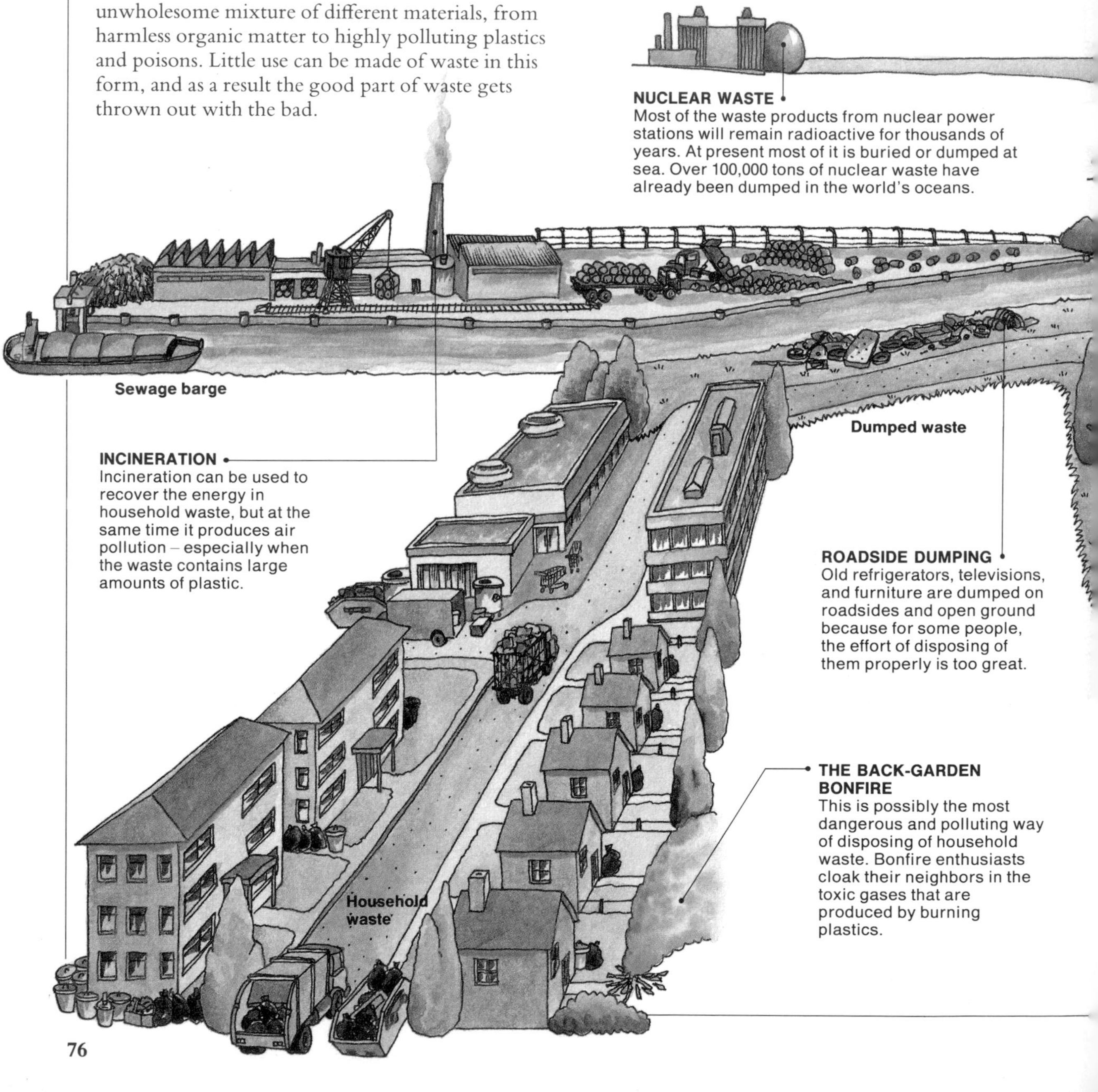

INCINERATION AT SEA
Incinerator ships burn chemical waste at high temperatures before releasing it into the atmosphere. This pollutes the sea, but because it is cheaper than treating waste on land, it continues on a large scale.

Nuclear power station

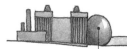

NUCLEAR WASTE
Most of the waste products from nuclear power stations will remain radioactive for thousands of years. At present most of it is buried or dumped at sea. Over 100,000 tons of nuclear waste have already been dumped in the world's oceans.

Sewage barge

Dumped waste

INCINERATION
Incineration can be used to recover the energy in household waste, but at the same time it produces air pollution – especially when the waste contains large amounts of plastic.

ROADSIDE DUMPING
Old refrigerators, televisions, and furniture are dumped on roadsides and open ground because for some people, the effort of disposing of them properly is too great.

THE BACK-GARDEN BONFIRE
This is possibly the most dangerous and polluting way of disposing of household waste. Bonfire enthusiasts cloak their neighbors in the toxic gases that are produced by burning plastics.

Household waste

DUMPING AT SEA
The sea – a traditional dumping ground – is increasingly used as a repository for sewage and industrial waste. Dumping one environment's waste on another is the worst way of disposing of it.

LANDFILL SITE
In most countries, by far the largest proportion of household garbage is buried. Once covered over, the waste pollutes water supplies and the surrounding soil. Toxic buried waste may also poison deep-rooted plants.

INDUSTRIAL WASTE
The Western world annually produces about 1 billion tons of industrial waste. In the United States nearly 3 million tons of concentrated acid and 2.5 million tons of solvents are thrown away every year.

waste will have rotted down, vast amounts of plastic, glass, and metal will still remain as evidence of our day-to-day lives.

Plastic will be more abundant than any other material. Although much of it will have been broken and torn, it won't have disintegrated. The names of shops will still be visible on plastic bags. Toy cars, plastic plates, spoons, razors, hair rollers, yogurt containers, transistor radios, and kitchen scales will still be identifiable, although the metal parts will have rusted away. If the people who uncover these things are fortunate enough to live in a more enlightened age, they will probably wonder *why* – quite suddenly in human history – we started to produce such huge masses of waste. They will want to know how it was that we came to discard up to ten times our body weight in the form of textiles, paper, glass, plastic, metal, and kitchen waste every year.

THE RISE OF CONSUMERISM

The garbage mountain began to grow during the great economic boom of the post-war era, a period of prosperity which brought with it a total change in production and consumption patterns, and also in people's attitudes towards throwing things away. It started in America. Self-service, which was fast replacing old-fashioned personal service, needed goods that were highly packaged. At the same time, marketing experts were trying out all kinds of new tricks to get people to buy more things more often, or to "stimulate consumption," as the economists would have it.

In the mid-fifties the marketing consultant Victor Lebow wrote a plea in the New York *Journal of Retailing* for "forced consumption." He said that "our enormously productive economy . . . demands that we make consumption a way of life, that we convert the buying and use of goods into rituals, that we seek our spiritual satisfactions in consumption . . . . We need things consumed, burned up, worn out, replaced, and discarded at an ever-growing rate." This view was echoed by the chief economist in the world's largest advertising agency of the time, J. Walter Thompson. He asserted that in 1960 Americans would have to learn to expand their personal consumption by 16 billion

A week's waste
The garbage in this household trash can (*left*) makes up a small-scale ecological disaster area. It is typical of the waste produced by millions of homes every week of the year. Potentially useful organic waste is mixed up with large quantities of plastic. Paper and cardboard, that might have been recycled, has also been tossed in, while useful raw materials, like aluminum foil, metal cans, and bottles, complete the mixture.

dollars a year if they were to keep pace with production ability.

And this is just what happened. The tail began to wag the dog: production produced consumption. Industrialists, trade unionists, wholesalers, retailers, marketing men, and politicians were all interested in a consumer boom which could keep the conveyor belts moving by quickening the pace of consumption. The creation of waste was quite deliberate. Americans were gradually turned into insatiable consumers of manufactured products. And what happened in America was mirrored in the rest of the western world.

OUT OF SIGHT AND OUT OF MIND

One of the best ways to encourage people to produce garbage at home is to take it away from them as fast as they accumulate it. For if you don't see the results of living wastefully (and you will not, because for the most part the garbage mountain is discreetly hidden from the public gaze), you will not be troubled by the rate at which you produce waste. The waste collection service that most of us enjoy doesn't just take garbage away – it creates more of it.

No wonder we give the trash collectors a tip when they come to collect the garbage in the last days of December. Never are the trash cans as full as in the week after Christmas. To get rid of

HOW MUCH WASTE WE THROW AWAY

This chart shows how much household waste is produced in a range of countries that all share high rates of consumption. The figures show the amount of waste that is actually collected every year. They do not take into account the waste that is dumped on roadsides and open spaces.

	Annual domestic waste (tons)	Equivalent per person (pounds)
United States	200,000,000	**1,930**
Australia	10,000,000	**1,500**
Canada	12,600,000	**1,157**
New Zealand	1,528,000	**1,075**
Norway	1,700,000	**915**
Denmark	2,046,000	**880**
Netherlands	5,400,000	**840**
Japan	40,225,000	**758**
West Germany	20,780,000	**742**
Switzerland	2,146,000	**740**
Belgium	3,082,000	**690**
Sweden	2,500,000	**661**
Finland	1,200,000	**639**
France	15,500,000	**635**
Great Britain	15,816,000	**622**
Italy	14,041,000	**542**
Spain	8,028,000	**472**

High-rise garbage
Through a cloud of scavenging seagulls (*right*) the buildings of Newark, New Jersey, rise up over the city's garbage dump. The inhabitants of cities produce far greater levels of waste than those in rural areas. Although the average annual waste produced per person in the United States is just over 1,900 pounds, someone living in a city may produce well over a ton.

that much extra waste requires a special inducement. Now if all the polystyrene packaging, wrapping paper, turkey bones, wine bottles, Coca-cola cans, plastic containers for fruit and vegetables, cardboard boxes, used-up batteries, old newspapers, and *last* year's toys were to remain on our doorsteps, we might think twice about throwing so much away, and therefore about buying so much potential waste in the first place. But it is all taken away for us, so we don't have to worry about it.

But what happens to our garbage once it has been carted away? In most countries the bulk of it ends up in a garbage dump, or landfill site as they are also known. A visit to one of these gives a striking picture of what we throw away. As well as huge volumes of packaging, rotting organic waste, broken bottles, cans, worn-out furniture, and household appliances, there are lesser quantities of much more dangerous materials – part-filled containers of cleaning fluids, pills, pesticides, wood preservatives, paints, paint strippers, glues, nail varnish removers, used oil from cars, and the corroding remains of old batteries. As all this gets dumped, it is crushed together in one foul-smelling, unmanageable and poisonous mass.

Household waste is an indiscriminate mixture of a huge variety of things which ought never to have been thrown in the same waste bin in the first place.

Once this mixture has been covered over by earth, it starts to react with itself. The organic matter, whose proper place is of course on a compost heap, starts to rot down. It produces an inflammable gas which makes its way to the surface (in some dumps this is piped off and used for heating). Substances from the chemical part of the garbage – mercury, cadmium, and nickel from batteries, waste solvents, weedkillers, and the like – are washed through the soil to reappear in drinking water.

And all this garbage takes up a huge area of land. All too often small valleys and potentially useful farmland are swallowed up by it, and just the thinnest layer of topsoil is applied to hide it away. The land is unsafe to build on and will not support healthy crops. But the garbage is out of *sight*, which is all that seems to matter.

ENERGY FROM WASTE

What happens when there is no space left to dump the garbage? Well, faced with this problem a growing number of authorities charged with getting rid of waste have decided to burn it. Until recently, waste incinerators were the brave new solution to the garbage problem. And why not? To get rid of waste by burning it and to produce energy at the same time seems both economically attractive and ecologically sound.

But as more and more waste incinerators were

built, the alarm bells started ringing. Big waste incinerators do not only get rid of garbage, they also release some highly dubious and dangerous gases into the air. If the temperature in the incinerator is below 1,650°F (as it often is), plastics, pesticides, and wood preservatives can give off dioxins as they are burned, and these are among the most poisonous substances known to science.

One of these, TCDD, is the dioxin which made its debut as a notable poison when it was unleashed from a chemical factory at Seveso in Italy, contaminating the ground so effectively that it is still unsafe years later. Even in tiny amounts it causes a variety of health problems, the best-known being chloracne, a disfigurement affecting the face, neck and shoulders. This can take two years to clear up.

If an incinerator is kept well flushed with oxygen, and if the garbage in it is stirred up well enough, the level of TCDD in the incinerator smoke can be kept down to a minimum. But so far, few incinerators are equipped with the necessary technology, and anyhow, no-one really knows if this "minimum" is safe. The absence of international standards of emission control has left many incinerator operators just guessing about the safety measures they should take.

THE DANGERS OF BURNING GARBAGE

Dioxins are just some of hundreds of ingredients that make up incinerator smoke. In it too is hydrochloric acid from plastics, as well as cadmium, lead, mercury, and selenium, although a considerable proportion of these metals remains in the ash which has to be dumped on hazardous waste sites.

Useful though they may sound, energy-from-waste incinerators produce more heavy metal pollutants for each unit of useful energy generated than any other plant fired by solid fuel. If our garbage is to be burned with any degree of safety, a lot of money has to be spent undoing the results of our thoughtlessness. Because our waste is such a jumble of different materials, a host of highly expensive filters and scrubbers are needed to clean up the smoke it produces.

Well, if you can't bury waste, or burn it, you can always dump it at sea. Many countries already do this: but the sea is not limitless, and it is already

THE GARBAGE-RICH HOME

Household garbage comes in two main types – the leftovers from day-to-day living, such as packaging, bottles, cans, and newspapers, and then once-in-a-while waste such as broken appliances, old furniture, worn-out carpets, and so on. The latter is an inescapable result of modern life, but the former, which makes up the bulk of household garbage, can be greatly reduced without too much difficulty.

POSITIVE ACTION

Six steps for reducing your household garbage

● **Don't mix up your garbage**
Ideally, every house should have a separate can for glass, paper, metal, and organic matter so that each can be recycled. Recycling kitchen waste is not possible if you don't have a garden, but recycling paper and glass certainly should be.

● **Apply the overpackaging test**
Excessive packaging is the most avoidable source of household waste. Choose products that are contained in the least amount of packaging, and never buy those that fail the overpackaging test (see p. 84).

● **Buy in bulk**
Regular household products packaged in small quantities produce more waste than those packaged in large quantities. Six separate cans of drink, for example, will produce far more waste than a single bottle containing the same amount. So whenever possible, buy the biggest size.

● **Choose returnable containers**
Returnable bottles are vastly better than cans, although over recent years they have greatly declined in number. Choose them and other returnable containers whenever possible.

● **Choose natural packaging**
Packaging made of paper or cardboard is preferable to plastic because it can be recycled. For the same reason, glass bottles are much better than plastic ones – especially if they are returnable.

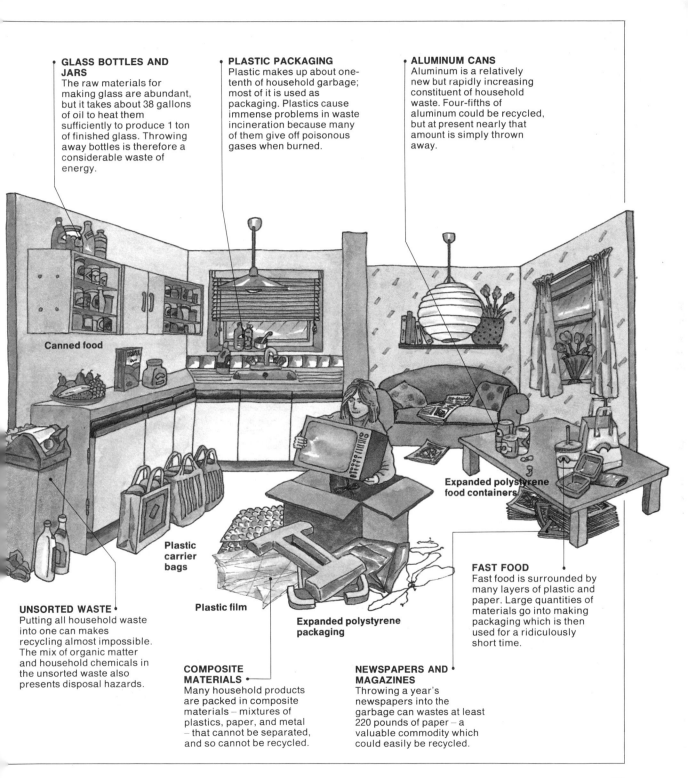

GLASS BOTTLES AND JARS
The raw materials for making glass are abundant, but it takes about 38 gallons of oil to heat them sufficiently to produce 1 ton of finished glass. Throwing away bottles is therefore a considerable waste of energy.

PLASTIC PACKAGING
Plastic makes up about one-tenth of household garbage; most of it is used as packaging. Plastics cause immense problems in waste incineration because many of them give off poisonous gases when burned.

ALUMINUM CANS
Aluminum is a relatively new but rapidly increasing constituent of household waste. Four-fifths of aluminum could be recycled, but at present nearly that amount is simply thrown away.

Canned food

Plastic carrier bags

Plastic film

Expanded polystyrene packaging

Expanded polystyrene food containers

UNSORTED WASTE
Putting all household waste into one can makes recycling almost impossible. The mix of organic matter and household chemicals in the unsorted waste also presents disposal hazards.

COMPOSITE MATERIALS
Many household products are packed in composite materials – mixtures of plastics, paper, and metal – that cannot be separated, and so cannot be recycled.

NEWSPAPERS AND MAGAZINES
Throwing a year's newspapers into the garbage can wastes at least 220 pounds of paper – a valuable commodity which could easily be recycled.

FAST FOOD
Fast food is surrounded by many layers of plastic and paper. Large quantities of materials go into making packaging which is then used for a ridiculously short time.

Energy from waste The world's largest household waste incinerator is at Ivry sur Seine near Paris. Here giant grapnels deal with 50 tons of refuse an hour as it is brought into the plant by the city's refuse trucks.

beginning to suffer from the burden of poisons and garbage that are every day being forced on it.

FROM WASTE TO COMPOST

As the garbage problem has become ever more intractable, a growing number of cities and towns have felt that they have to do *something* useful with it rather than let it pile up in ever-larger mounds. Why not turn it into compost?

One such composting works was built at Siggerwiesen, outside Salzburg in Austria. It was designed to process the household waste of half a million people. Increasing living standards and the ever-growing number of tourists who come to visit this magnificent city of Mozart, and the beautiful countryside around it, have also meant huge piles of garbage, in fact over 100,000 tons of it each year. The city's waste disposal planners set about trying to turn this rapidly growing liability into an asset.

At Siggerwiesen household refuse is delivered in garbage trucks which discharge their loads into deep pits on the outside wall of the composting plant. Kitchen waste, newspapers, cardboard, metal cans, glass, batteries, and plastic are all dumped together. The waste is ground up in crushing machines, and heavy hammers pulverize glass and flatten metal cans. Iron particles are removed by magnetic separators, while some of the plastics and non-ferrous metals are extracted. The waste is transferred to fermentation drums, each of which is the size of a couple of steam engines, and at this stage in goes sludge from the nearby sewage works.

As the drums slowly rotate, and as air is pumped into them, the garbage starts to ferment and the decomposing brew heats up. After two days at a temperature of up to 170°F, most of the dangerous germs in the garbage have been destroyed. By now the mixture has started to become compost. It is spread out in storage halls where air is pumped through it, and there it stays for six to eight months. During this time the waste breaks down into mature compost whose volume is only one-fifth of the garbage that made it up. During this final "ripening," all the remaining disease-causing germs die. The compost is then fully matured and ready for use.

The purpose of composting organic household wastes and sewage sludge is, of course, to capture their inherent fertility and return it to the land. Compost is excellent for enriching the soil with organic matter and it also creates ideal living conditions for the myriad earthworms, soil insects, microbes and fungi that live in the top 12 inches of the soil, just below our feet.

THE PROBLEMS OF COMPOSTING WASTE

So is the Siggerwiesen type of composting plant the solution to urban waste problems worldwide? Well, nearly but not quite. Again, the spanner in the works is the huge variety of waste that has to be dealt with.

As we have seen, household garbage is dangerous stuff. It contains heavy metals from batteries, and even arsenic from disinfectants and pesticides. Some of these can be removed from the garbage as it is sifted before fermentation, but some pro-

WHAT IS IN HOUSEHOLD WASTE?

The average household throws away about two cans of garbage a week, creating an annual pile that weighs 2 tons or more. If all this remained at home instead of being carried away, every house would be surrounded by cans full of rotting waste, rusting metal, discarded packaging, bottles, broken household appliances, and seeping chemicals. As a mixture, this waste is completely useless.

THE RECYCLING POTENTIAL OF HOUSEHOLD WASTE

The diagrams below show how household waste is made up. They also show what potential each of the ingredients has for being recycled. A huge proportion of household garbage – glass, paper, metal, and organic matter – is fully recyclable. This part of household garbage accounts for about 80 percent of its total weight. The remaining 20 percent of waste is composed of plastics, solid and liquid chemicals, and composite materials, including packaging and mixed-fiber clothing. These are difficult to recycle, although with techniques such as those described on pages 87–9, some of the plastic could be reused.

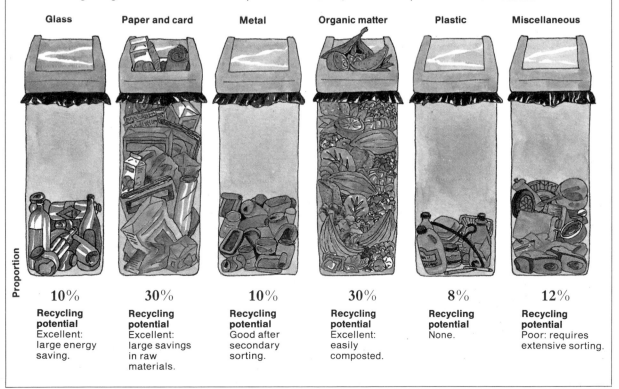

Glass	Paper and card	Metal	Organic matter	Plastic	Miscellaneous
10%	**30%**	**10%**	**30%**	**8%**	**12%**
Recycling potential Excellent: large energy saving.	**Recycling potential** Excellent: large savings in raw materials.	**Recycling potential** Good after secondary sorting.	**Recycling potential** Excellent: easily composted.	**Recycling potential** None.	**Recycling potential** Poor: requires extensive sorting.

Proportion

portion of them will get into the compost. The sewage sludge, too, contains heavy metals from households, factories, and farms. So the compost, however rich in plant nutrients it may be, is not without problems on the chemical front. To make it less attractive still, bits of plastic invariably end up in the mixture, although they are very small.

Now, since farmers have heard and read any number of news stories about heavy metal contamination of soils in recent years, they are reluctant customers for compost which contains such impurities. They know that they may find it difficult to sell farm produce – vegetables, grain, fruit, meat, or milk – which contains heavy metals. So Siggerwiesen compost is sold mostly to flower-growers and for use in parks and vineyards. It may also find a use in forests with soils that are poor in nutrients. But it isn't used for growing food.

So composting the great mish-mash of waste that makes up household garbage is clearly not the answer. What we need to do is to *reduce* the amount of waste and *recycle* what waste is still produced. And that means sorting it when it is thrown away, so it never gets mixed up in the first place.

RECYCLING ORGANIC WASTE

If you live in the country and have a garden, organic matter should never be part of the waste that you put out for collection. If it is, it is your loss, because vegetable peelings, leftovers from meals, and all other kitchen scraps are potentially very valuable. There is no need to get anyone to take them away to compost them: they should be kept at home and used on the garden.

However, if you live in a city and do not have a garden, it is clearly not practical to compost your own kitchen scraps. In this case it does have to be taken away, and so it needs to be set aside from inorganic waste. This is a very simple business. In several towns in Germany, homes have two trash cans, a gray one and a green one. These are emptied alternately. Kitchen waste, together with paper and cardboard, is put into the green can, and when this is taken away, the contents go straight to a composting yard where they are rotted down for use as farm or garden compost. The garbage in the gray trash can is taken to the dump or incinerator.

POSITIVE ACTION
Garbage recycling in the garden

- **Kitchen leftovers**
  All kitchen waste is useful in the garden – even eggshells and bones, which both provide valuable plant nutrients.

- **Natural-fiber clothing**
  Wool, cotton, and linen will all rot down quickly if buried. Wool, in particular, is a good source of nitrogen which promotes plant growth.

- **Newspapers and cardboard**
  These will decompose, albeit fairly slowly, if buried in damp soil or put on a compost heap.

- **Cellulose-based products**
  Cellulose wallpaper paste (pure – *not* with fungicide), cellophane, and cellulose filler will all decompose in the soil.

THE OVERPACKAGING TEST

A revealing way to examine packaging is to count up the number of separate layers that surround whatever it is that you are buying. There are few products that really need more than two layers of packaging, but many have far more than that. Luxury items are the worst offenders: the less essential a product is, the more it will be wrapped up – as can be seen here with a box of chocolates.

Seven-layer packaging

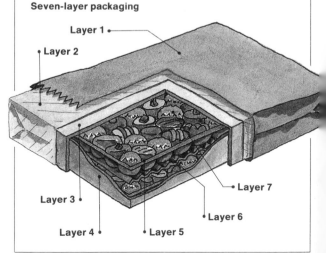

Packaging pollution
Overflowing waste bins, like this one in Japan, are becoming more and more common as packaging becomes increasingly elaborate. Packaging companies are moving away from traditional materials like waxed card and are replacing them with foamed plastics and aluminum. The result is great masses of lightweight garbage which swamps trash cans and which blows about in the wind.

This is only a partial sorting system, but it does produce a compost that is considerably safer than the all-in-one-can method. The extra brainwork it requires is minimal: there cannot be many people in the world who would be unable to decide into which of two cans their garbage should go.

AVOIDING OVERPACKAGING

So recycling organic matter should be quite straightforward. But what about all the packaging that we bring into our homes? What can individuals do to help reduce the great mountains of it that are piling up all over our Earth? Well, first we must be aware of the problem. Then everybody needs to keep the problem in mind as they go about their daily business.

Anyone who can remember the English countryside of three score years ago will recall how housewives used to take small calico bags with them when they went to the village shop. These would have, neatly embroidered on them, the names of the things they were to contain: sugar, tea, coffee, rice, flour, and so on. The shopkeeper kept all such substances in bulk, in jute sacks, wooden bins, pottery crocks, or galvanized iron containers. He would weigh the required amount of each substance in his scales and pour it straight into the customer's calico bag (plastic, of course, had not yet been invented). This simple method of shopping not only eliminated the need to cut down forests to make paper bags, it also avoided any waste packaging being created.

Today, about one-tenth of the average weekly shopping bill is spent on packaging alone. Much of this serves no real purpose. If enough people refused just the superfluous bags that are thrust on them by shop assistants, this waste would be avoided.

But packaging is not only used to pack things – it is also used as a means of making sure that a customer has paid for whatever he or she is carrying out of a shop. But if a shopkeeper will not trust you to walk to the door with an unwrapped purchase, go to another shop.

Another nonsense that should be rejected is the so-called "individual" pack. The more an item is divided up, the more packaging, in total, is required to wrap up the same amount of a substance. There is no need for this. The individual pats of stale butter wrapped in metal foil, which are served up so often in restaurants and snack bars, are an example of this method of producing extra waste. In energy and garbage disposal the wrapping costs far more than the butter inside. Similarly, every single sugar-cube is wrapped in a piece of paper in the interests of "hygiene" and "convenience". But what's wrong with sugar in a bowl?

It is not only food that gets this treatment. Why should we pander to the silly habit of manufacturers who wrap six little brass screws in an elaborate container and sell them for six times the price of similar screws that a hardware dealer can sell twisted up in a piece of newspaper? Individual packs only clutter up our planet with waste.

THE PROBLEM OF PLASTIC

When Leo Baekeland invented Bakelite in 1909, he could hardly have realized what a profound effect this and other plastics were going to have on the future of the Earth. For the very virtues that make plastics so useful – their strength and their durability – are precisely what make them such an appalling problem.

The trouble with plastics is that they are put to all sorts of uses for which they are actually quite inappropriate. We use plastics, which are long-lasting materials, to make objects that we promptly throw away.

Let us take a shopping-bag as an example. Now the material that the shopping bag is made from will last for centuries: this makes the bag virtually immortal. But how long is the bag's useful life? An hour? Perhaps two? In return for using up oil in its production, and creating something that will litter the Earth long after we have gone, we have an object that gives us two hours of service.

Now that may sound bad enough. But many plastic objects have useful lives that are measured not in hours, but in minutes or even *seconds*. Foam

Waterborne garbage
Not that long ago, all human garbage that found its way to the sea rotted quickly and sank. Now most inshore waters are littered with plastic (*above*) caught up in the ebb and flow of the tides.

Wildlife hazards
Aquatic animals can get trapped by submerged plastic waste. This seal (*below left*) is caught in a discarded fishing net; a trout (*below right*) has grown inside a plastic ring.

POSITIVE ACTION
How to cut down on waste plastic

- **Don't buy fresh food prepacked in plastic**
  In many supermarkets, meat and vegetables are prepacked in plastic, to the ridiculous extent that four tomatoes may have their own plastic tray surrounded by plastic film. This is a quite unnecessary source of plastic pollution. Whenever possible avoid prepacked food.

- **Refuse plastic containers for fast food**
  You don't need them. Most of these containers are thrown away within minutes. Paper bags are all that is required.

- **Choose natural materials in clothing**
  Avoid man-made fibers in clothing, even those present as mixtures with natural materials. Also, if you can, avoid the plastic that is disguised as leather in shoes.

- **Avoid disposable plastic in the kitchen**
  Plastic is long-lasting, so it makes no sense to use it in objects that are disposable. Use plastic food boxes instead of throwaway film and bags.

- **Ask for biodegradable packaging**
  Your local supermarket manager may not have heard about biodegradable plastic bags (see p. 88). Try suggesting that they use them instead of permanent plastic.

public water supply. Even in the home, PVC, which is widely used in floor tiles, coat hangers, shower curtains, records, plastic bottles, margarine containers, and food wrapping film, gives off vinyl chloride gas. Trace quantities of this gas, which has been found to cause liver cancer in workers employed in plastics factories, are commonly found in households. But far larger quantities are given off by garbage dumps — dumps which are rarely far from towns and cities.

Most garbage dumps catch fire at one time or another and these can burn for weeks or even months. PVC and polystyrene foams are usually involved. Anyone who has breathed the smoke from these two substances will know how acrid it is: it contains hydrogen fluoride, which is not only very poisonous but also contributes to acid rain.

So getting rid of plastics is not at all easy. It makes far more sense not to use so much of them. But as long as we use gasoline in our cars plastics will be with us because they are made from naphtha, a by-product of gasoline refining. So the sooner we learn to use them more prudently, and find some way of *reusing* them, the better.

RECYCLING PLASTIC

Although all plastics may look very similar, chemically they may be as different from one another as chalk and cheese. Some of them can be melted down, while others, if subjected to the same process, will simply stay solid until they catch fire, which is not much help to anyone. In the past all the processes designed to make something useful out of plastic waste have foundered because of this.

But eventually it may be possible to reuse some of this mountain of plastic waste. Machines do exist that can recycle all the plastics we use in our households. In this process, the plastics that melt when heated, and that make up about three-quarters of the whole, are turned into a sort of sticky dough. The plastics that remain solid are then added like raisins in a cake mix. The mix can be poured into molds and turned into new products such as heavy-duty tiles for factory floors, pallets, sewage pipes, railway sleepers, or fence posts.

The value of this process is that it can make use of all plastic waste, from "clean" offcuts from plastics

polystyrene food-containers of the kind used by fast-food shops are brought into this world to keep food warm for as long as it takes to carry it out of a shop. Then it is into a trash can with it, or more often, on to the pavement. Like the shopping bag, the container's life is potentially limitless. But what is created for a useful life of the utmost brevity will remain with us more or less forever. The same is true of all plastic food packaging. Plastic film, plastic boxes, plastic bags — they all end up in the garbage as soon as the food is taken out.

WHAT HAPPENS TO WASTE PLASTIC

Well, what if we do use so much plastic? It can always be buried or burned, can't it? Well yes it can, but at a price.

When plastic is buried, that is not always the end of the matter. Red and yellow plastics usually contain cadmium as a pigment. Rain can carry this into groundwater and from there it gets into the factories to dirty plastic extracted from garbage dumps. The raw materials in the process are, of course, usually free, apart from the cost of extraction and transportation, and they can be used to make objects that would otherwise have required new materials. The energy needed to make this recycled plastic is only a fraction of that needed for producing aluminum, steel, or new plastic.

PLASTICS THAT ROT AWAY

If the plastics that make up so much of packaging could be deprived of their immortality, much of the garbage mountain would disappear without anyone being the worse for it. It doesn't matter at all if a material only has a life of a year or so if it is only being used for a day.

It so happens that at the moment nearly all the

plastics we use are chemically very hardy: they completely resist any form of biological decomposition. But plastics do not have to be like this. It is possible to create materials that have all the good qualities of plastics, but which can be broken down by bacteria in compost heaps – just as if they were paper, cardboard, or kitchen waste.

At present nearly all plastics are made from oil. Biodegradable plastics, however, are made by the fermentation of natural substances such as sugar and other carbohydrates. One firm has produced biodegradable plastic with the help of a vigorous strain of bacteria found in canals. The bacteria are cultivated in vats and fed a sugary diet on which they thrive. In doing so they multiply and produce a biological "plastic" rather like mammals make fat in their bodies as they grow. The plastic is extracted

WHAT HAPPENS TO DISCARDED GARBAGE?

In the last two decades litter from cans and bottles has become almost universal. What happens to this waste after it has been thrown away? Will natural processes get rid of it? The answer depends on the material.

Metal cans and glass bottles eventually break up fairly harmlessly, but plastic bottles, an ever larger part of packaging, are resistant to all forms of natural chemical decay and are practically indestructible.

ALUMINUM
Aluminum in cans reacts with oxygen in the atmosphere, but forms a layer of oxide which protects it from decomposition. Aluminum garbage takes many years to disintegrate.

  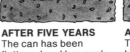 

AFTER ONE YEAR Most of the paint has dissolved, but otherwise the can is intact.

AFTER FIVE YEARS The can has been flattened and has sunk into the soil.

AFTER TEN YEARS The can is very slowly being decomposed by contact with the soil.

GLASS
Glass is a harmless and wholly inert substance – it does not chemically decompose at all. However, it does disintegrate, although after burial this process more or less stops.

AFTER ONE YEAR The bottle is still intact on the surface.

AFTER FIVE YEARS The glass has broken into large fragments.

AFTER TEN YEARS The glass, now in small fragments, lies buried harmlessly in the soil.

PLASTIC
Many plastics are broken down to some extent by ultraviolet light: this clouds them and makes them brittle. However, once underground most buried plastics do not decompose at all.

 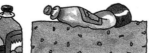  

AFTER ONE YEAR The bottle is in much the same state as when it was thrown away.

AFTER FIVE YEARS Sunlight has partially decomposed the plastic, but the bottle is intact.

AFTER TEN YEARS Once buried, the plastic will remain intact almost indefinitely.

POSITIVE ACTION

How to step up glass recycling

- **Help to make recycling more efficient**
  Removing contaminants such as bottle tops and plastic stoppers makes the recycling process more cost-effective, because less sorting is then needed after collection.

- **Ask for a bottle bank**
  If you haven't got a bottle bank near you, ask for one. Glass manufacturers are usually interested to learn of new sites for collection.

- **Don't waste returnable bottles**
  Your enthusiasm for recycling glass should not extend to throwing returnable bottles in a bottle bank: reusing bottles saves far more energy than melting them down.

Recycling glass After waste glass has been broken up into manageable fragments, any metal and plastic caps are removed before the glass enters the furnace.

in fermentation vessels, and is then dried and sold as granules. This plastic is readily broken down by algae, fungi, or bacteria in the soil. A bag made from it will disappear within twelve or fifteen months or indeed within only three or four months if it is placed on a compost heap.

At present, the cost of producing plastics like this is far greater than the cost of ordinary plastics. But the reason for this is that it is carried out on a small scale. "Permanent" plastic should only be used when permanence is really needed. Biodegradable plastic should be the material of first choice. When this happens, its cost will drop rapidly, and plastic litter will disappear. If enough shoppers simply refused to have any dealings with packaging made from permanent plastic, the problem would be solved quickly enough.

SAVING GLASS

No one could claim that the raw materials from which glass is made are in short supply. There is enough silica on Earth to make untold billions of bottles, if we could find the necessary energy to melt it down. And therein lies the reason for saving all your empty bottles and jars: to make a bottle requires a great deal of heat, which must ultimately come from a limited supply of oil.

Glass makes up about one-tenth of household garbage and is one of the easiest of materials to recycle. All you need to do is take it to a bottle-

bank. The broken glass, or "cullet" as it is known, which is gathered in these banks, is then added to new molten glass in a furnace. The result is new glass produced with considerably less energy than would otherwise be needed.

Even less energy is needed if bottles don't have to be melted down at all, but are reused instead. It is time that the returnable bottle staged a comeback. In 1972, the state of Oregon in the United States passed a "bottle bill" which required all drinks containers to carry a deposit, and so to be returnable. Because bottles are the easiest containers to reuse, many manufacturers reverted to using them instead of plastic or metal. Other states have followed suit, and any country with an ounce of interest in preventing litter and conserving energy should surely do the same. Choosing returnable bottles is the best way to help this happen.

RECYCLING ALUMINUM

The environmental price paid to make an aluminum can is a horrifying one. Aluminum is one of the most expensive and polluting metals to produce: the energy needed to make one soda can is the equivalent of half that can filled with oil. The metal is extracted from bauxite ore which is mined on the surface, much of it in tropical forest areas. This mining process destroys large areas of natural vegetation and with it, tropical insects, birds, and mammals. Mining leaves the soil bare and erosion then causes rivers to silt up, depriving fish of their habitats, and depriving fishermen of their livelihood. Huge quantities of slag are left behind on sites where alumina has been produced from bauxite. The alumina then has to be treated in a series of chemical processes to produce the aluminum metal. All sorts of pollutants, including fluorine gas, are released on the way.

Melting down an aluminum can so that it can be reused requires just 5 percent of the energy needed to make a new one. It creates no pollution at all. So every can that is thrown away is a lost opportunity to save energy and preserve the environment.

Once again, all that is needed is intelligent sorting of garbage, or a system for returning a used can in exchange for a new one. Aluminum cans can be collected by vending machines that work in reverse: instead of putting money in, and getting a can in return, you put in the can and take the money.

Of course, aluminum cans should never be used in the first place. Digging up tropical forests in order to extract bauxite and produce aluminum which is then sent all over the world contradicts just about every principle of ecologically sound living that can be thought of. Every country has abundant raw materials for making glass, and it is this eminently recyclable material that should be used to hold drinks.

RECYCLING PAPER

Most of the paper that we use has a useful life of a few weeks, or, in the case of newspapers, less than a day. At present, less than a quarter of the paper in use is recycled. But there is nothing to stop *all* paper being recycled, except perhaps our obsession with appearances.

Most people are accustomed to using new paper which is white as snow. The reason that it is white is not because this is paper's natural colour, but because it is bleached. The bleaches used in paper-making are one of the major environmental hazards of that industry, polluting rivers and lakes in quite appalling ways. Recycled paper, on the other hand, which is just as good as new paper, tends to be slightly grainy and gray or light green. It does not get this thorough bleaching, so as well as saving timber, it also reduces water pollution.

Until recently most paper was made from rags but today virtually all of it is produced from wood pulp. It has been established that recycling half of the paper used in the world today would meet almost three-quarters of the demand for paper and as a result free 10 million acres of forest from paper production.

So why is this not done? Simply because of laziness, and because paper that is less than brilliant white is somehow less acceptable. For this luxury, vast areas of the world's surface are given over to producing a substance in quantities that greatly exceed our real need.

Paper is one of the easiest materials to recycle. Every home should have its paper-store which is emptied regularly into a paper-bank. In addition to this, there should be an exchange system for newspapers. Every time a newspaper is sold, an old newspaper should be required in return. If this happened, the amount of timber needed for newsprint would plummet overnight. Paper recycling, which is at present only of marginal financial interest, would become hugely profitable, and the scourge of paper litter would vanish.

MIXED MATERIALS

How do you sort garbage when some of it is made of mixtures of materials, or materials glued together? The answer is you can't. Now this is clearly infuriating to anyone who wants to recycle their household waste – but what does the packaging industry think of the problem? Not a lot, if the current vogue for composite materials is anything to go by.

Packaging, of course, has two main purposes: to provide a suitable container for a product which it is meant to protect and to preserve, and to make the product look attractive to potential buyers. Impressive looks have always been a criterion for the "acceptability" of consumer goods, but never more so than today. Composite materials are a boom area for the packagers: they see nothing but a rosy future for packaging made out of such mixtures as aluminum foil and cardboard, polythene and aluminum foil, polypropylene and paper. Unfortunately, these are all combinations which are impossible to recycle.

RECYCLING HOUSEHOLD WASTE

The four main ingredients of household garbage, organic matter, glass, paper, and metal, are all easily recycled. Recycling materials has a threefold benefit. It reduces the problem of waste, it reduces the amount of energy needed to make household products, and it reduces the pollution and destruction that getting hold of new raw materials so often causes.

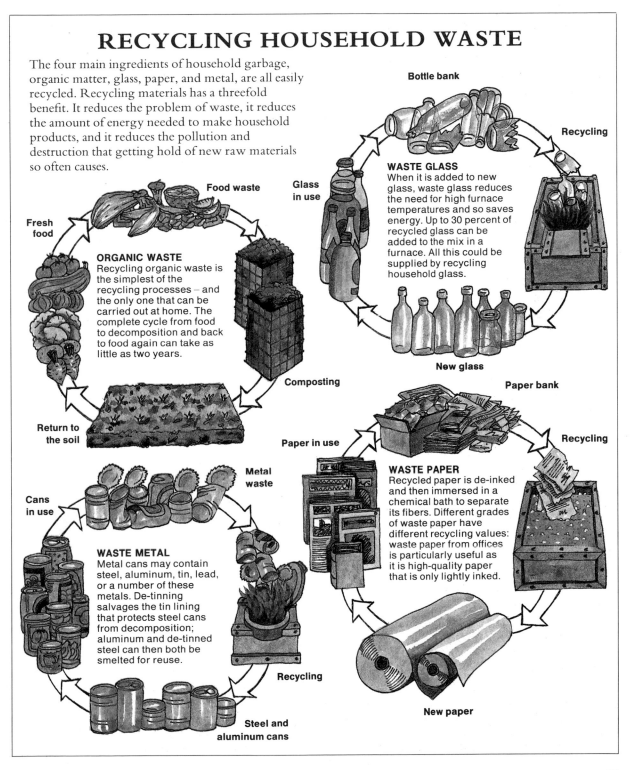

Bottle bank

Recycling

Glass in use

WASTE GLASS
When it is added to new glass, waste glass reduces the need for high furnace temperatures and so saves energy. Up to 30 percent of recycled glass can be added to the mix in a furnace. All this could be supplied by recycling household glass.

New glass

Food waste

Fresh food

ORGANIC WASTE
Recycling organic waste is the simplest of the recycling processes – and the only one that can be carried out at home. The complete cycle from food to decomposition and back to food again can take as little as two years.

Return to the soil

Composting

Paper bank

Paper in use

Recycling

WASTE PAPER
Recycled paper is de-inked and then immersed in a chemical bath to separate its fibers. Different grades of waste paper have different recycling values: waste paper from offices is particularly useful as it is high-quality paper that is only lightly inked.

Metal waste

Cans in use

WASTE METAL
Metal cans may contain steel, aluminum, tin, lead, or a number of these metals. De-tinning salvages the tin lining that protects steel cans from decomposition; aluminum and de-tinned steel can then both be smelted for reuse.

Recycling

New paper

Steel and aluminum cans

"The snack industry sets a hectic pace in innovation, development, and new product launches", enthuses the glossy publicity for a leading brand of composite packaging. "Success in this highly demanding sector is dependent on product and point-of-sale appeal. The brilliance of metallized polypropylene films and laminates ... enhances colorful graphics and their barrier performance gives longer shelf-life." And what of their potential as garbage? Nothing: no mention that such materials are quite impossible to recycle.

Composite packaging can neither be recycled as metal, paper, nor plastic. The different constituents cannot be separated and they cannot easily be broken down by microbes in a garbage dump. As more composite packaging materials are used, it will become increasingly difficult to sort your household refuse should you wish to do so.

What is needed is a list of ingredients on each product which also says something about the *packaging*. A product wrapped in composite materials should carry an environmental health warning, advertising the fact that it contains non-recyclable materials which will create pollution. Everyone would then know what to avoid, and packagers might have second thoughts about using these mixtures.

GARBAGE FROM INDUSTRY

Much of the poisonous garbage produced by factories is the by-product of making the ordinary things we use in our daily lives. When this waste is dumped it can grow to spectacular and horrifying dimensions. We have already heard about waterborne waste in places like Love Canal and the Niagara River. Solid waste on land is just as much of a problem. In Georgswerder near Hamburg in West Germany there is one especially notorious industrial dump which contains 200 million cubic yards of waste which has been dumped on the site since 1948. The garbage mountain is 45 yards high: you can see it clearly as you drive along.

The history of Georgswerder shows the total lack of foresight typical of most industrial waste disposal. From 1967 to 1974 it was the final resting place for up to 40,000 cubic yards of toxic industrial waste a year. A potent mixture of highly poisonous chemical wastes, industrial sludges, and old engine oil was dumped in shallow basins lined with a thin layer of plastic. The arch-poison TCDD seeped out of the dump into local fishponds and groundwater. Virtually all the dangerous chemical effluents imaginable are to be found in the dump.

Safe disposal of hazardous industrial wastes has become a huge problem for all industrialized countries. The Georgswerder dump is now being sealed by surrounding it with a concrete wall and covering it with a layer of clay to prevent the penetration of rainwater (no mean feat considering that it covers as much ground as a large village).

The industrial wastes that are not disposed of at dumps, rivers, or the sea are burned in special incineration plants. They shed hydrochloric acid, dioxins, and heavy metals into the air which are then eventually deposited on forests and farmland. Acid rain releases these pollutants in the soil. Much farmland is now so polluted with heavy metals that it ought no longer to be used to produce food for human consumption.

While the use of gasoline has fallen slightly, the consumption of heavy metals is still going up. Cadmium and lead may remain in the environment for long periods. So far no method has been found for removing them once they have accumulated in the soil. Any further releases of these substances from factories, incinerators, or garbage dumps will do lasting damage to the living world and with it the health of our children.

We can do quite well without these poisons, and no doubt we would do without them if we were told which products generated them. Left to itself, industry will go on producing toxic materials that, no matter how they are gotten rid of, will boomerang back on us. But if we can eventually have labels that itemize both ingredients and packaging, let us finally have one more piece of information: itemized toxic waste. Then we can decide exactly what chemical garbage we wish to participate in producing.

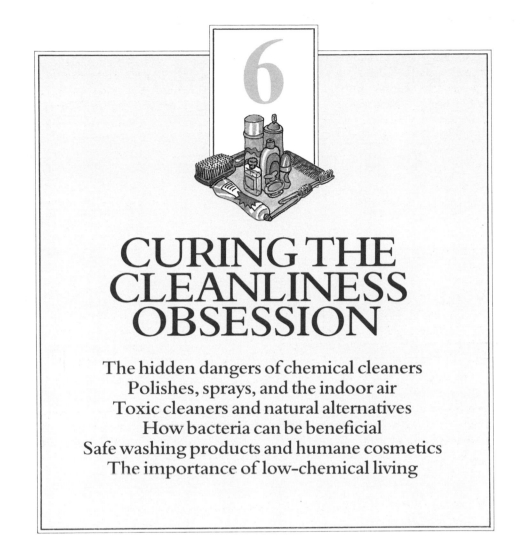

6

CURING THE CLEANLINESS OBSESSION

The hidden dangers of chemical cleaners
Polishes, sprays, and the indoor air
Toxic cleaners and natural alternatives
How bacteria can be beneficial
Safe washing products and humane cosmetics
The importance of low-chemical living

When, in the 1860s, Joseph Lister in England, and others elsewhere in the world, collectively invented the "germ theory," they probably did not know that they were starting a multibillion dollar industry and also, in some ways, doing a great deal of harm.

Lister put his theory into practice by carrying out major operations on patients while their wounds were being sprayed constantly with strong carbolic acid. Any germs that were around at the time died, but alas, so did a good many of the patients – not from the germs but from the carbolic acid. Later, less ruthless methods of providing asepsis were invented, and no doubt many lives have been saved by them. But Lister and his colleagues inadver-

tently started a phobia among the people of the Western World. This "germ phobia" has gotten right out of control and is now heavily exploited by people out to make money by preying on the ill-founded fears of others.

The average home is the scene of a seemingly endless battle to eliminate germs, but the germs simply will not be eliminated. Our houses are cleaner, tidier, and shinier than ever before, but this is thanks to the use of substances that make the world outside a worse place. Not only that but we have become convinced that our bodies, too, have to be germ-free. Again, this involves the use of polluting chemicals. Of course there is no such thing as a germ-free, odorless body, nor will there

THE COST OF CLEANLINESS
How chemical cleaners affect the environment

All too often keeping somewhere clean simply shifts pollution from one place to another. Nearly all the chemical cleaners used in the home involve the production of polluting substances. Great play is made of their effectiveness, but no mention is made of what their use does to the environment.

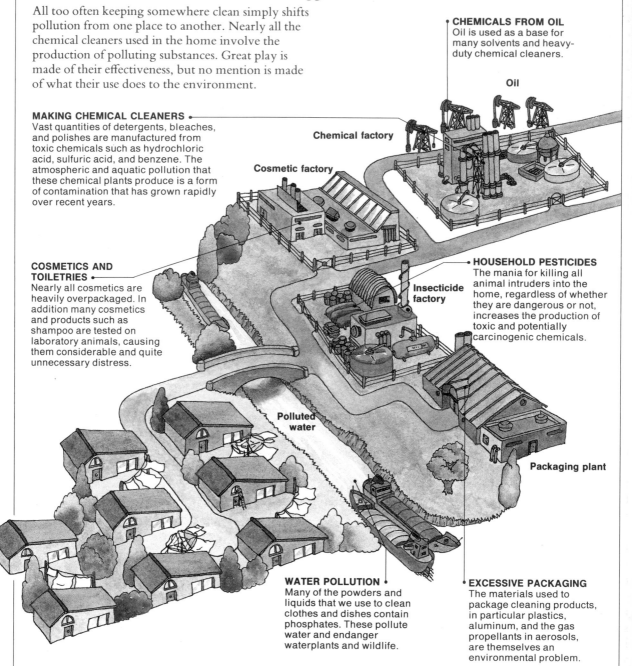

CHEMICALS FROM OIL
Oil is used as a base for many solvents and heavy-duty chemical cleaners.

Oil

Chemical factory

MAKING CHEMICAL CLEANERS
Vast quantities of detergents, bleaches, and polishes are manufactured from toxic chemicals such as hydrochloric acid, sulfuric acid, and benzene. The atmospheric and aquatic pollution that these chemical plants produce is a form of contamination that has grown rapidly over recent years.

Cosmetic factory

COSMETICS AND TOILETRIES
Nearly all cosmetics are heavily overpackaged. In addition many cosmetics and products such as shampoo are tested on laboratory animals, causing them considerable and quite unnecessary distress.

HOUSEHOLD PESTICIDES
The mania for killing all animal intruders into the home, regardless of whether they are dangerous or not, increases the production of toxic and potentially carcinogenic chemicals.

Insecticide factory

Polluted water

Packaging plant

WATER POLLUTION
Many of the powders and liquids that we use to clean clothes and dishes contain phosphates. These pollute water and endanger waterplants and wildlife.

EXCESSIVE PACKAGING
The materials used to package cleaning products, in particular plastics, aluminum, and the gas propellants in aerosols, are themselves an environmental problem.

ever be, but the advertisers persuade us that if we strive to develop one, then we will have no shortage of friends, fame, or fortune in life.

The men and women of sixteenth-century Europe covered everything in lashings of perfume to mask the considerable odors that resulted from dirty clothes and bodies – and from the sewage which used to run in stinking rivulets down the middle of the roads. They had little interest in washing or cleaning. That is a situation few people would care to return to. But today the pendulum has swung in the other direction. We are not just hygienic (a good thing), we are compulsive cleaners (a very bad one). Where a scouring brush and a strong elbow would once have been sufficient, there is now an army of chemical cleaners waiting to be brought to the frontlines of the hygiene battlefield.

CHEMICAL CLEANERS IN THE HOME

Liquid and powdered bleaches are some of the most polluting household chemicals. They often feature at the scene of the householder's longest-running engagement with germs – the toilet. Millions of horrible germs are lurking round your toilet bowl. Or so the advertisements tell us. And the implication is that although they will not actually leap out and bite us, nevertheless they may well infect us with some horrible disease. Bleach, we are told, is the answer to these horrors.

Bleach is a very unpleasant substance. It does more than simply kill bacteria in the toilet – it keeps on killing them long after it has been flushed away. Whatever system of sewerage your house may have, it must ultimately work by bacterial action, and bacteria cannot act if subjected to constant dousing with virulent poisons. Disinfectants are likely to interfere with the action of the benevolent bacteria that digest our sewage for us and render it harmless while large amounts of bleach may harm septic tanks. So great care is needed – and moderation.

These unwanted effects on the sewerage system are quite apart from the health risks that using something as poisonous as bleach presents in the home. Bleach on its own is dangerous enough if you touch it with bare hands, or if you are young enough to confuse it with a bottled drink. But it is even worse when mixed with other household chemicals. Combining different cleaners can be very tempting with so many products on the market. One doesn't seem to be doing its job properly, so why not try using another one too? It will strengthen the cleaning power, won't it? Well, no, not always: if bleach is mixed with an acidic cleaner, chlorine gas can start forming within a matter of seconds. This gas can build up very quickly and can be fatal.

The toilet is not the only thing we clean so obsessively. We also wash our dishes and clothes with great regularity. As we have already seen in chapter 2, dishwashing liquid, dishwasher and laundry detergents, all of which are used and sold in such huge quantities, should be treated with caution.

Detergents are a very recent invention on this planet, and are at present being grossly abused. Nearly all of them contain phosphates. They all go down the drain, and not only is this a terrrible waste of phosphates, but it causes great damage to water life. The rivers and lakes of industrial countries, and the plants and creatures that live in them, are being destroyed by this excessive phosphate build-up.

To watch the average person dumping detergents into dishwater you would think that he or she was being paid a bonus for the amount used. The recommended amounts to be put in washing machines are about double those really needed – after all, the recommenders are trying to sell the stuff. They are hardly likely to err on the low side as to quantity. But effective detergents are being made nowadays that do not have phosphates in them at all. If you use detergents, these are the ones to choose. Whatever detergents you use, you need not use so much.

As with all man-made chemicals, detergents bring health risks with them. If they are swallowed, dishwashing liquids cause nausea, vomiting, and diarrhea. The drying agent which is designed to remain on the surface of the crockery creates a film on it so that the water slides off more quickly. But this substance may increase the body's absorption of DDT and other pesticides present in food, and the harmful effect of these pesticides on our health is becoming increasingly apparent.

POLISHES, SPRAYS, AND THE INDOOR AIR

Floors, windows, and furniture come in for a lot of cleaning, and many of the substances designed to do the job are packed in aerosol canisters.

The time saved by using an aerosol, compared with the polishes out of an ordinary can, is minimal, while the manufacture of aerosols uses up scarce materials. They are not recyclable and are therefore sheer waste. But aerosols also threaten the health of the user and of the planet as a whole. Their propellants (the pressurized gases that make them work) are suspected of either polluting the atmosphere and destroying the protective ozone layer, or of causing cancer, depending on which sort of gas is used in the propellant.

There is simply no justification for using aerosols: they should be banned completely and immediately, but in default of such banning each one of us should simply make the decision not to use another one under any circumstances. They are a dangerous gimmick without which the world got on perfectly well for thousands of years and it is high time we kicked the aerosol habit for good.

Synthetic polishes, by their very smell, warn us that they are not healthy. Like polishes, many other products contain artificial perfumes to disguise their ingredients, but some are even designed to release these perfumes to give a house what advertisers call a "fresh" smell. But no one should have any illusions about this so-called freshness. Any house that smells strongly of these chemicals does so, not because it is fresh or clean, but because the air in it is polluted.

Deliberately contaminating household air with artificial perfumes is bad enough, but contaminating it with poisons is pure folly. No one who cares at all for their health should have anything to do with airborne insecticides, especially the vapor-issuing cake that is hung up to ward off flies. It certainly kills flies, but any chemical that is toxic to one form of life is almost certain to be harmful to other forms too. It is better to be bothered by flies than to poison yourself. There is a simple device, although rather expensive, that electrocutes flies by attracting them with an ultraviolet light. It is highly effective and far safer than using any of the airborne poisons that are available.

HOUSEHOLD CLEANING WITHOUT CHEMICALS

Any animal living in a fixed and permanent den, as most humans do, has to be concerned about cleanliness, for if it is not, it soon becomes infested with parasites. Anyone who has lived in hot and unsanitary countries and has come into contact (close contact) with such things as bedbugs will appreciate the reason for this. But all unwelcome visitors in the home can be easily eliminated by simple cleaning routines, instead of by using virulent substances.

Sweeping, brushing, beating carpets, washing floors, and laundering sheets and clothes and blankets, will keep the home adequately free of pests, especially if food is stored out of reach. Vacuum cleaners are also an efficient means of keeping the place clean. The invention of the vacuum cleaner has put the flea circus ringmasters out of a job – they can no longer find actors! Vacuum cleaners make fairly moderate demands on the nonrenewable resources of the planet and do not consume too much energy.

There are many simple and safe alternatives to the household chemicals that poison our environ-

POSITIVE ACTION

How to minimize pollution from household cleaners

- **Use natural products**
  Whenever possible, use products that contain natural, nontoxic ingredients. These are likely to be less harmful to your health and the environment.

- **Use moderate amounts**
  When using substances such as bleaches and detergents, be sparing with them. This will reduce water pollution.

- **Avoid hazardous packaging**
  Think carefully about how your cleaning products have been packaged. Cardboard detergent boxes can be recycled, but products in aerosol cans or plastic bottles are difficult to dispose of and damaging to the environment.

- **Read labels carefully**
  If a product is labeled as potentially toxic, think twice before buying it: use a harmless alternative instead (see opposite).

ment and ourselves while claiming to make both more acceptable. Simply using less of them is a good way to begin. But even to take such straight-forward action you first have to arm yourself against a battery of propaganda aimed at us by people who dispose of a large part of the money there is in this world. An important step in the quest for cleanliness lies in developing a resistance not to

germs, but to advertising: this would solve half the problem at a stroke.

Let's start with the toilet. If you are really worried by the aesthetics of a slight encrustation of lime and other minerals in your lavatory basin, you can get rid of it by neutralizing the lime with vinegar and then quite simply brushing it off. Common sense will suggest that the vinegar

NATURAL ALTERNATIVES TO CHEMICAL CLEANERS

Chemical cleaners are formulated to give instant results, and are therefore often both highly concentrated and potentially harmful. But you don't have to use synthetic chemical cleaners. Traditional alternatives, based on natural products, will do the job just as adequately. This table outlines the problems with five types of chemical cleaners and suggests natural alternatives for each.

Product	Chemical cleaner	Natural alternative
Toilet cleaner	Chemical toilet cleaners often contain bleach which in turn contains sodium hypochlorite. This is a highly caustic agent which pollutes water and destroys the bacterial balance in sewage.	A strong solution of a natural acid, such as vinegar, will remove most limescale without causing water pollution.
Laundry detergent	Synthetic detergents pollute water and play havoc with the skin. Many of their ingredients, such as perfume, have no practical value at all.	For hand-washing, soap and small amounts of washing soda dissolved in hot water make an effective cleaner. For washing machines, phosphate-free powders will reduce water pollution.
Dishwashing liquid	Many dishwashing liquids contain phosphates which are highly damaging to water life. Detergents dissolve fats rapidly, and, along with them, the skin's natural oils.	In soft-water areas, hot water and soap will remove grease effectively. Soda and soap dissolved in boiling water work well on more ingrained dirt.
Furniture polish	Most synthetic furniture polishes are based on synthetic silicones and solvents. They often contain artificial perfumes which pollute the indoor air.	A natural furniture polish can be made by combining 2 parts olive or vegetable oil with 1 part lemon juice. Beer, sugar and beeswax is another alternative.
Metal polish	Metal cleaners often contain ammonia, which can burn the skin, and petroleum distillates which are highly poisonous if swallowed. They may also be hazardous when inhaled.	Aluminum foil in a salt solution will remove tarnish from silver. Lemon juice will clean brass and copper, and apple cider vinegar will clean chrome.

solution must be strong and left in for some time. (Obviously if you can siphon the water out of the bowl and put the vinegar in neat it will work better, but this probably isn't something most people would care to try!)

As far as the dishes and clothes are concerned, it is the hot water and the physical action of washing that does a large part of the cleaning. What hot water cannot deal with unaided can be removed with nothing more complicated than soap. Warm water and soap will remove all grease from any article.

Making a nonpolluting detergent is quite easy. If you dissolve bar soap and washing soda in boiling water, you will make an excellent soft soap. This is perfect for washing either crockery, cutlery, or clothes, and will also make a good natural aphid-killer for the garden (see p. 135).

Soft soap is ideal for hand washing clothes, but will probably not work in your washing machine. If you must go on using detergents (and being so used to them you probably must) use non-phosphate ones. The principles of responsibility and moderation should come into force here. You should know what you are doing, what you are using, and what effect it is having. And you should be moderate in all three.

It is hard to see why anyone needs synthetic polishes. There are plenty of good old-fashioned polishes available, so why go in for newly invented chemicals which may do you harm? Beeswax is a traditional and perfectly harmless way of polishing wood: what artificial polish could boast of such a healthy production process as that carried out by bees in a hive?

In general, aim to use cleaning products that are as close to their natural state as possible. If necessary you can make them yourself by combining substances like oil or lemon juice or vinegar, which break down naturally when they have done their job. If a child gets hold of a bottle of lemon juice it might feel rather ill, but it will not be likely to die as it might after trying out some scouring powder or disinfectant. And without the factories that turn out these household chemicals, the Earth would be a cleaner and safer place. Without the chemicals themselves our homes would be much safer too.

CLEANLINESS AND COEXISTENCE

Bacteria, viruses, and microscopic fungi are around us all the time – we are covered with them, full of them, and surrounded by them. But not all micro-organisms cause disease. Countless millions of bacteria live on our skin and in our digestive systems. Even potentially harmful bacteria such as those that cause tuberculosis are found in the lungs of the healthiest people. Our immune system keeps them at bay. If we eat good wholesome food and not too much of it, exercise hard and frequently in the open air, none of these micro-organisms will harm us because there are such things as healthy bacteria in healthy bodies. There is a natural coexistence between all the forms of life on this planet.

Nowadays it seems many people are unable to tolerate the idea that they must live with these tiny but harmless organisms. They have been fooled into believing that the constant use of bathroom chemicals will make them germ-free. We are told that the Queen of Sheba and Cleopatra always bathed in milk to soften and whiten their skin, while Mary Queen of Scots bathed in either asses' milk or wine. Whether or not these ablutionary habits had any benefits for the skin, we cannot know, but one thing is sure: they were certainly less harmful to their health than many of the things we smear, splash, and spray over our bodies today.

THE DANGERS OF DEODORANTS

Deodorants work by suppressing or killing the micro-organisms that live on sweat. Although some of these bacteria can be harmful, it is important to establish a balance, but deodorants destroy all the bacterial flora and leave the skin open to attack from germs that can in turn cause severe skin irritations.

The substances that are designed to kill these micro-organisms and bacteria are very similar, in the effects they have, to the insecticides that are used on plants by the less responsible farmers and gardeners of this world. As with plants, disturbing the natural balance of organisms does not eradicate them, it simply creates havoc as the multiplication of certain species goes unchecked by others.

Antiperspirants, which are often combined with deodorants in a single product, have a different function: they reduce the amount of sweat on the skin. Humans do not pant like dogs: we remain cool by the evaporation of sweat. Any chemical that inhibits the body's mechanism for controlling its temperature must be regarded with some suspicion. Washing with soap and water works as efficiently in the short term and has none of the long-term hazards of deodorants or antiperspirants.

The containers that these deodorizing products so often come in are, once again, the infamous aerosols. As we have already seen, using these is an act of irresponsibility that everyone should try to avoid committing.

WHAT'S WRONG WITH SOAP?

Soap, as it used to be, was a relatively harmless substance. For hundreds of years it was made entirely from natural substances – the main ingredient being animal fat. But today's soaps are different, because the marketing men have moved in and encouraged the manufacturers to start adding what they call "selling points". We use soap, and other simple little products such as toothpaste, without considering that they might be crammed full of chemicals of one sort or another. But they are, and many of them are not at all good

for us. Additives to the basic soap now make it smell "nice" or prevent it cracking and drying out, while artificial colors allow us to have soap to match the bathroom decor.

Research has raised questions about the harmful effect of modern soaps on the skin's protective layer, but this has not stopped the manufacturers introducing yet more products such as shower gels and soaps with built-in deodorants. Extravagant claims are made about their benefits, but there is no evidence that any of them are an improvement on traditional soap. Some, like deodorant soaps, might actually be worse: they contain substances that collect on the upper layer of the skin and which can sometimes get into the bloodstream, or cause allergic reactions.

That is not all that gets into soap. Hexachlorophane is a bactericide often used in soaps and deodorants, as well as many other hygiene products. In 1972 a baby powder appeared in France which contained this chemical. When it was used in a Paris hospital it led to the death of 36 babies, while 150 children suffered from long-term effects. Investigations revealed that this particular product contained ten times the recommended amount of hexachlorophane due to a manufacturing error,

Inundated with aerosols
Since aerosols first appeared in the late 1940s, endless rows of them have become a familiar sight in every supermarket. The cost to the environment of keeping millions of houses clean, perfumed, and polished with aerosol products is enormous. As well as using up metals and polluting the atmosphere with their propellants, they explode and release dangerous gases when incinerated.

and that it could be extremely dangerous even in small quantities. In West Germany this chemical is banned in baby products, and its concentrations must be clearly labelled on other products.

Despite the experience with hexachlorophane, new chemicals are continually being added to soap. For the sake of health and environment, they are best left well alone. Soap in its original form requires no modern "improvements".

NATURAL HAIR CARE

Many of the shampoos sold today claim to have extraordinary properties: they can make your hair blond or black or auburn – or silver or blue for that matter; they can make your hair grow faster and thicker and longer; and they can make it shine and curl. Well, they certainly *claim* to do these things.

It has been said that hope is what is sold in cosmetics. If that is so, the shampoo-buying public seems to be very hopeful, because little good comes of any of the new formulations – they are almost always nothing but marketing ploys. Even if a shampoo contains an ingredient that in its natural state and in large amounts *could* effect small changes in your hair, the quantities of the necessary ingredient are likely to be negligible in any mass-produced shampoos on the market.

Of all the shampoos sold, one-tenth are said to combat dandruff, to which end they contain such substances as selenium sulphide, which is sometimes used in batteries. Yet these supposed cures have only a temporary effect. The customer is always having to go back for more, so the consumption of polluting and quite unnecessary raw materials continues to rise.

Shampoos were traditionally made from combinations of water and a range of natural ingredients such as herbs, vinegar, fruit juices, egg yolks, and even beer. They were gentle and effective. Modern washing agents, by comparison, remove dirt but they also take about four-fifths of the hair's natural oils with it. In addition, some of the colors and preservatives added to today's shampoos and hair-dyes irritate the skin and eyes. They have in some cases been linked with increased incidence of cancer: chemicals are absorbed by the scalp and then work their way into the blood and urine.

THE CHEMICALLY CLEANED HOME

Each year the average family goes through enough chemical cleaners to fill a large-size bath. Most of these synthetic substances damage the environment, either in their manufacture or through their effect on the water cycle after they have been used. In addition to this, a number of them cause indoor pollution. The air you breathe can be adversely affected by a variety of chemicals leaked into the indoor atmosphere by cleaning products.

POSITIVE ACTION

How to keep clean without using chemicals

Thousands of toiletries and cosmetics jostle for attention in chemists, drug stores, and supermarkets. How do you choose the ones that are best for you and best for the environment? These guidelines will help.

● **Look for natural ingredients**
Disregard all reckless claims about revolutionary man-made ingredients and instead stick to products that are known to be safe – those made from natural ingredients.

● **The simpler the better**
Plain soap and plain shampoo are just as good as their colored equivalents. Useless additional ingredients simply create more pollution from chemical factories.

● **Apply the overpackaging test**
Whenever possible, buy whatever is contained in the minimum amount of packaging. Cosmetics in particular are ridiculously overpacked: when buying, apply the test shown on page 84.

● **Avoid deodorants**
These disrupt and damage the skin's bacterial flora. Try not to buy them.

● **Cut down on quantity**
We are not as smelly or as dirty as advertisers would have us believe. Help your body by cutting down on the quantity of chemicals that you use on it.

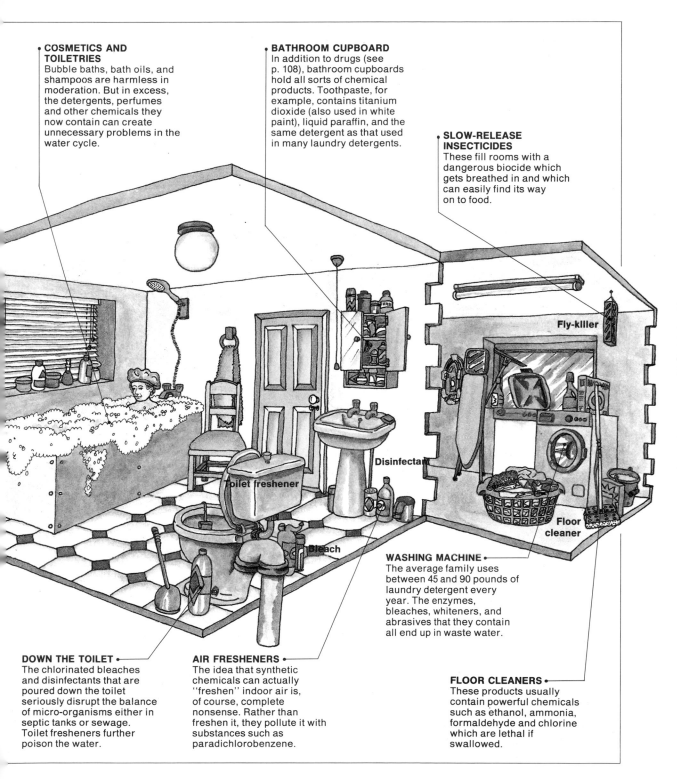

COSMETICS AND TOILETRIES
Bubble baths, bath oils, and shampoos are harmless in moderation. But in excess, the detergents, perfumes and other chemicals they now contain can create unnecessary problems in the water cycle.

BATHROOM CUPBOARD
In addition to drugs (see p. 108), bathroom cupboards hold all sorts of chemical products. Toothpaste, for example, contains titanium dioxide (also used in white paint), liquid paraffin, and the same detergent as that used in many laundry detergents.

SLOW-RELEASE INSECTICIDES
These fill rooms with a dangerous biocide which gets breathed in and which can easily find its way on to food.

Fly-killer

Disinfectant

Toilet freshener

Bleach

Floor cleaner

DOWN THE TOILET
The chlorinated bleaches and disinfectants that are poured down the toilet seriously disrupt the balance of micro-organisms either in septic tanks or sewage. Toilet fresheners further poison the water.

AIR FRESHENERS
The idea that synthetic chemicals can actually "freshen" indoor air is, of course, complete nonsense. Rather than freshen it, they pollute it with substances such as paradichlorobenzene.

WASHING MACHINE
The average family uses between 45 and 90 pounds of laundry detergent every year. The enzymes, bleaches, whiteners, and abrasives that they contain all end up in waste water.

FLOOR CLEANERS
These products usually contain powerful chemicals such as ethanol, ammonia, formaldehyde and chlorine which are lethal if swallowed.

101

Yet it is completely unnecessary for anyone to put themselves at risk by using these potentially dangerous products. Many companies are now producing shampoos made entirely from natural ingredients, and the original elements like lemon and vinegar are hardly unobtainable. If all synthetic shampoos disappeared overnight, we would all be the better for it.

HOW ANIMALS SUFFER FOR OUR VANITY

You cannot make yourself truly beautiful by smearing cosmetics on your face, but you can create an illusion of a sort of beauty. People nowadays are all subjected to a constant barrage of propaganda which makes them want to look like those models with perfect physiques and flawless complexions. In purely commercial terms, this propaganda is highly successful. The average American uses between 10 and 45 pounds of soaps, toiletries, and cosmetics every year in an attempt to achieve this perfection – an appalling waste of time, money and energy.

Cosmetics are, of course, quite unnecessary, but there is no point being too puritanical about them: self-adornment is not a habit that we humans are likely to give up. However, there is no doubt at all that the finest cosmetic available in the world is brisk hard manual work or exercise in the fresh air!

But although some cosmetics may be quite harmless, either to the environment or to health, the principle of nonviolence should surely make us loath to use cosmetics that have been tested on animals. Every year, thousands of rabbits are subjected to the Draize Test and many more are used in other tests on raw materials for the cosmetic industry. The Draize Test consists of putting the substance to be tested into the eyes of a live rabbit and then watching to see if there is inflammation and formation of pus.

An even more unacceptable practice is the notorious LD50 (Lethal Dose 50) test – an experiment that determines how much of a substance is needed to kill half of the animals it is given to. It has been shown that the results of this experiment varied by a factor of eight between different laboratories when testing the same material. Surely such a test, even with a substance that could be a matter of life or death, is unjustifiable?

Fortunately, there is a growing reaction to the excessive use of animals in the testing of cosmetics, and more and more makers of cosmetics are now prepared to guarantee that their products have not involved any form of cruelty to animals.

In England the first branch of the Body Shop chain opened in 1976 with a strict policy of only making and selling such products. The company has a five-point philosophy which should become a charter for all cosmetics manufacturers: first, that animals should not suffer in laboratory tests for human vanity; second, that whales should not be slaughtered in order to make moisture creams; third, that placenta should not be used in cosmetics; fourth, that aerosols are an unacceptable form of packaging because of their damaging effect on the environment and because they are indisposable; fifth, that the packaging should be inexpensive to buy and economical to produce. The phenomenal success of the company (it now has outlets from Iceland to Singapore) has shown how much feeling there is about the subject of cosmetics and animal suffering.

LOW-CHEMICAL LIVING

If we are to survive for many more generations on this planet, we will have to develop an organic approach. We will have to realize that we are part of a mighty and diverse biotic community. We must co-exist with other species, large and small – even microscopic. The dream that we can have perfect control over our environment – that we can eliminate every other form of life on this planet except the forms of life that are directly useful to us – is a bad dream and a dream that will never come true for us. If we really want to achieve that, then we must go and live on the moon.

It does not add to the quality of our lives for us to be dirty, or for our houses to be infested with beetles or cockroaches, mice or rats. Although we have to coexist with other forms of life on this planet, we still have a right to stay clean and to live in homes that are reasonably free from these creatures. But as Uncle Toby said in *Tristram Shandy*, as he caught the fly that had been bothering him and put it, alive, out of the window: "There is room in the world for thee and me!"

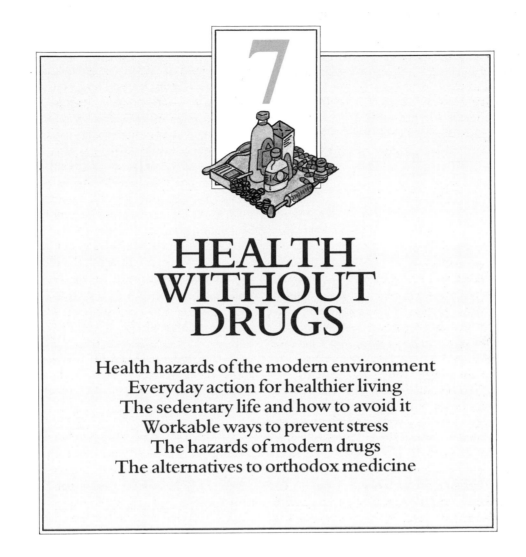

7

HEALTH WITHOUT DRUGS

Health hazards of the modern environment
Everyday action for healthier living
The sedentary life and how to avoid it
Workable ways to prevent stress
The hazards of modern drugs
The alternatives to orthodox medicine

The dream of an everlasting life in perfect health is probably as old as mankind. Today the average lifespan may be longer than ever before, but no one could honestly claim to have found an elixir that will prevent our bodies from aging, or indeed even one that will simply prevent them from falling ill. Life expectancy has certainly gone up, but it now stubbornly refuses to rise any further.

There are a number of reasons why we are now likely to live longer than our forebears. Better sanitation has put a stop to regular epidemics of typhoid, cholera, and dysentery which were common in the nineteenth century. Vaccination has made bacterial and viral diseases such as diphtheria and polio far less of a threat, while smallpox, the first disease to be treated by vaccination, has vanished altogether. Few of us now have to worry about once-dreaded diseases like pneumonia, tuberculosis, and even influenza – the invention of antibiotics has either prevented them or curtailed their effects. With synthetic drugs at hand, surgery, accidents, and childbirth, which at one time would have ended in fatality, are no longer a major worry.

But during this century we have acquired a number of quite *new* diseases. Instead of epidemics of smallpox or influenza, we have epidemics of cancer and heart disease and most of us are likely to die from them. The vast sums spent on medical research across the world have not provided us with reliable cures because the researchers are only looking at what happens inside the body, rather than at what is happening all around it.

HEALTH AND THE ENVIRONMENT

In 1900 the world boasted just one city, London, which had a population of more than 5 million. But by 1980, London had been overtaken by fifteen other cities, nine of which now have populations of more than 10 million. In western Europe, North America and Australasia, two people in every three are living in the great cities.

What effect does the urban environment have on the body? Without doubt, a bad one. The air itself is certainly less healthy than that of less crowded surroundings. The smog that shrouds cities like Los Angeles and Tokyo still causes severe respiratory problems, despite attempts to reduce it. Clean air laws may have gotten rid of soot from city chimneys, but in many towns and cities, traffic fumes have simply taken its place. The gases that are released into the air by vehicle exhausts are nearly all poisonous and city-dwellers cannot help but breathe them in every day of their lives.

Where lead is still permitted in gasoline, this additional poison ebbs and flows around city streets. It interferes with the proper working of the nervous system, and although few people are unfortunate enough to be crippled by it, no one who lives in a town or city escapes a daily dose of this substance. Other airborne triggers of ill-health include tobacco smoke and chemical fumes. The more built-up the area, the worse it is.

Urban water, as we have seen in chapter 2, is not exactly pure. With millions of people crammed into small areas, drinking water has to be reprocessed again and again. As a result the pollutants in it build up over time. Furthermore, if you live in a city, your food is more likely to be processed and even the "fresh" fruit and vegetables that you buy are likely to be weeks old.

But unpleasant though all these environmental problems are, they cannot be held entirely responsible for our ills. The *way* we live, rather than where we live, is probably more important still.

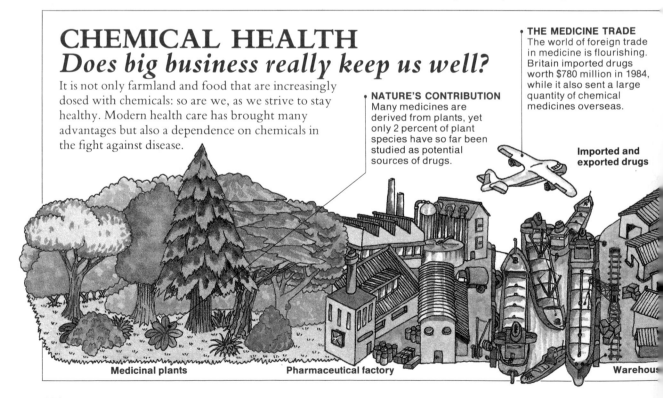

CHEMICAL HEALTH
Does big business really keep us well?

It is not only farmland and food that are increasingly dosed with chemicals: so are we, as we strive to stay healthy. Modern health care has brought many advantages but also a dependence on chemicals in the fight against disease.

THE MEDICINE TRADE
The world of foreign trade in medicine is flourishing. Britain imported drugs worth $780 million in 1984, while it also sent a large quantity of chemical medicines overseas.

NATURE'S CONTRIBUTION
Many medicines are derived from plants, yet only 2 percent of plant species have so far been studied as potential sources of drugs.

Imported and exported drugs

Medicinal plants   Pharmaceutical factory   Warehous

THE SEDENTARY LIFE

Stripped of all our fine clothes, our bodies are much the same as those of our hunter-gatherer ancestors. These people spent much of their time walking long distances barefoot, feeling the ground's many and varied surfaces with every step. Life for them was not particularly strenuous, but neither was it idle. For weeks at a time, families would wander a fair distance every day, trapping animals, collecting wild food, and setting up a temporary camp each night. They got a good deal of physical exercise.

In a modern sense, they were homeless. They had no permanent dwellings. Neither did they have many possessions, because all that they owned had to be carried from one camp to the next. But at the end of the day, once the food had been collected and eaten, there was little left to worry about. The few hunter-gatherers still left on this Earth, in such places as the Kalahari Desert, who have not been corrupted by modern living, are very healthy.

Things are very different for the rest of us. We are shut off from the outside world and encumbered with possessions. Our feet are encased in shoes and the only time their sensitive nerve endings receive any sort of stimulation is when we take our shoes off before going to bed. But far more important, we don't use our feet much any more. We don't walk, we drive. We don't use our hands for climbing rocks and trees or collecting food. Instead we use them for pressing buttons, flicking switches, or holding knives and forks. Far from wandering in the open, we spend our time indoors, giving our bodies little chance to do the kinds of activities they were designed for.

Physical exercise has become something that is only undertaken voluntarily – there is just no need for it in day-to-day life. Most people who do have a job sit at their desks doing overtime so that they can earn enough money to buy "labor-saving" devices, in other words machines that will reduce the need for physical activity still further.

THE EFFECTS OF STRESS

As anyone who has been pursued by a fierce dog will know, the human body, even a fairly unfit one, has within it a remarkable ability to react to sudden

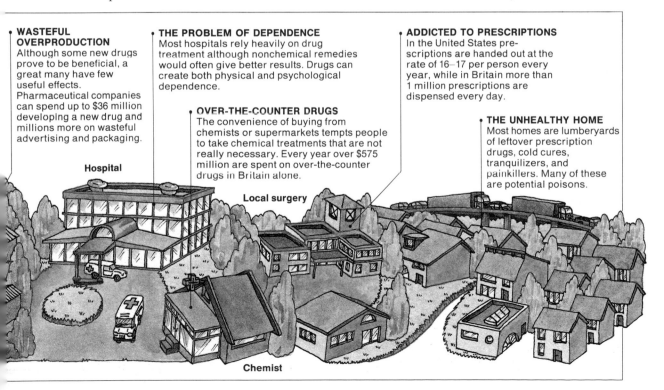

WASTEFUL OVERPRODUCTION Although some new drugs prove to be beneficial, a great many have few useful effects. Pharmaceutical companies can spend up to $36 million developing a new drug and millions more on wasteful advertising and packaging.

THE PROBLEM OF DEPENDENCE Most hospitals rely heavily on drug treatment although nonchemical remedies would often give better results. Drugs can create both physical and psychological dependence.

OVER-THE-COUNTER DRUGS The convenience of buying from chemists or supermarkets tempts people to take chemical treatments that are not really necessary. Every year over $575 million are spent on over-the-counter drugs in Britain alone.

ADDICTED TO PRESCRIPTIONS In the United States prescriptions are handed out at the rate of 16–17 per person every year, while in Britain more than 1 million prescriptions are dispensed every day.

THE UNHEALTHY HOME Most homes are lumberyards of leftover prescription drugs, cold cures, tranquilizers, and painkillers. Many of these are potential poisons.

Hospital

Local surgery

Chemist

danger. The heart pounds, the blood surges, and extraordinary reserves of strength and stamina are ready for the ensuing fight or flight. But the daily challenges of modern life are not as simple as this. They are abstract threats to which we cannot respond in a physical way. The body makes itself ready for action but is never called on to act.

The cumulative effect of threat without response is stress. The tyranny of the clock, trying to meet deadlines, rushing about to get to meetings or appointments, and hurrying meals all contribute to it. Whether you are stressed because of work, or stressed because you are out of it, the pressures of modern life are manifold – and they undoubtedly take their toll on health. The body is not able to cope with constant apprehension and it starts to show signs of distress. Digestive disorders, headaches, stomach ulcers, skin problems, depression, chronic lethargy, even cancer and heart disease, are all related to the mounting levels of stress in our lives.

With stress comes the search for an instant release from it. Alcohol is the traditional resort of those who for one reason or another cannot cope with the stress of daily life. Smoking is still very common, especially in young people, while glue-sniffing and drug-taking are almost endemic in cities. There are now 200,000 cocaine addicts in New York alone.

So what is the health profession doing about all this? How is it helping people to come to terms with the frustrations of life in apartment towers and crowded offices? Well, if you do fall ill and visit your doctor, the chances are that he or she will be as stressed as you are. Yours will be one of dozens of appointments that day, your problem will be assessed in a few minutes, and then you will be given a prescription. Pills and doctors have become inseparable. Much of the time, pills seem to make people feel better but this is because they relieve the symptoms rather than effecting a deeper cure. Doctors simply are not in a position to prescribe what is really needed – a different way of life.

HELPING YOURSELF TO HEALTH

Many full-blown diseases are signals by which the body complains about years of maltreatment. Tackling serious diseases may often require all the skill and technology of modern medicine, but with a bit of bodily "good housekeeping," a number of these diseases can be prevented in the first place.

The pollution we let loose on the world eventually finds its way back to our bodies and we suffer for it. What we do to the Earth, we do to ourselves and to those who come after us. We cannot damage the planet and get away with it and we cannot damage ourselves and expect to remain fit and healthy.

Now it is obviously impossible for us all to pack up a few belongings and then head out to the open country to live the life of hunter-gatherers. But we can do a great deal to improve our health by taking much better care of our bodies. For too long now they have been treated as if they were machines with changeable parts and inevitably short-lived efficiency. But we are not just a collection of spare parts. The heart is not simply a pump attached to pipes that must be unclogged from time to time with bypass operations, or else replaced when worn out. It is an instrument that needs looking after.

Our bodies are remarkable organisms, capable of

POSITIVE ACTION

Five steps for reducing stress

- **Put your health first**
  Combating stress means giving your health top priority. If you feel stressed, allowing your body to relax should be a regular daily activity.

- **Build up your natural defenses**
  Avoid stimulants and depressants such as tea, coffee, alcohol, and cigarettes. These adversely affect the nervous system, and can deplete the body's available supplies of vitamins and minerals.

- **Resist the lure of drugs**
  Chemical remedies cannot "cure" stress-related problems – they merely defer their effects. Avoid taking drugs for stress, whether they are antidepressants or tranquilizers.

- **Try natural tranquilizers**
  Many herbal teas have soothing properties. Camomile, passion flower and lime-blossom tea can all help to relieve stress and tension.

- **Get enough sleep**
  If you feel tired and run down, you will be easy prey to stress, so make sure you are getting as much sleep as *you* need.

endlessly regenerating themselves, healing cuts, grazes, bruises on the outside, and combating diseases on the inside. They show admirable resilience in the face of alcohol, cigarettes, lack of exercise, and poor nutrition. Yet if they were carefully and knowledgeably looked after, provided with the right food, sufficient exercise, and adequate amounts of rest, how much *more* efficiently they could operate.

Half of the rushing about that goes on today is quite unnecessary. We often compare ourselves to ants rushing about in their heaps. Well, at least ants have good reason to rush about; they are all doing important work. But most of our time we rush about whether we are doing important work or not. It becomes a habit. Simply slowing down is an important step towards preventing disease.

Exercising can not only help the body, it can improve the outside world. Walking or cycling instead of driving reduces pollution and makes life healthier for others. Even walking after a lawnmower is healthier than being carried around on one. Saving labor often means spoiling health.

Finally comes care about what you eat, for proper nutrition is essential for preventing disease.

THE IMPORTANCE OF DIET

Sir Robert McCarrison, a pioneer of medical research, spent about 30 years doing nutritional studies for the Indian Medical Service. In one set of experiments in 1926 he fed young healthy rats on three different diets. One group was fed on wholemeal cakes, pulses, raw vegetables, milk, butter, and a little meat – the traditional Sikh diet. A second group was fed largely on rice with some vegetables – the poor people's diet of southern India. The third group of rats was fed on the typical working-class diet in Britain at that time: white bread, margarine, canned meat, jam, sweet milky tea, and overcooked vegetables. The last group showed by far the worst health record. They had a high mortality rate, very few offspring, and severely damaged, inefficient digestive systems.

We are not rats in a laboratory experiment. We can choose the diet we wish to eat and we can also help determine the state of health we wish to enjoy. The average diet in the Western world now consists of a far better range of foods than white bread, margarine, and so on. But we are still influenced by our food. Choosing it carefully, and especially avoiding the kinds of processed foods described in chapter 4, will keep your body in a fit state to ward off most of the illnesses which now plague us.

THE RISE OF CHEMICAL MEDICINE

During this century, medical drugs have been discovered at an ever-increasing rate. Antibiotics, such as penicillin and streptomycin, the sulfa drugs, antihistamines, and antidepressants have all, in turn, been acclaimed as the longed-for remedy for one condition or another. And, let us make no mistake about it, some of them are wonder substances indeed.

But since the Second World War, the manufacture of these drugs has brought with it a massive growth in the pharmaceutical industry.

The chemical production-line Every day machines like this produce millions of capsules, tablets, and pills. The manufacture of synthetic drugs is a huge industry with an enormous output.

THE HOUSEHOLD DISPENSARY

Most homes contain enough dangerous drugs to kill their occupants many times over. Of course, all the pills, powders, creams, lotions, and syrups are used gradually, so their potential as poisons is rarely realized. But acute poisoning is just one hazard of having drugs in the home: they can have many other effects that are far from beneficial.

PAINKILLERS
The most widely used painkiller, aspirin, can damage the stomach lining and is dangerous for people with stomach ulcers, kidney problems, or high blood pressure. Over $70 million are spent on painkillers every year in Britain.

CONTRACEPTIVE PILL
The contraceptive pill, used by around 50 million women worldwide, is almost completely effective. However, apart from a number of minor side-effects, it may be linked with thrombosis, strokes, and certain types of cancer.

ANTACIDS
Antacids relieve the effects of bad diet, poor eating habits, and tension. In Britain $37 million worth of antacids are used each year. Although they reduce acidity, they can also cause digestive problems.

LAXATIVES
As with any drug that affects the digestive system, laxatives can cause disorders themselves, and they can also put great strain on the kidneys. Every year in Britain $36 million are spent on them.

TRANQUILIZERS
One in five people will take tranquilizers at some time in their life, not realizing the high risk of dependency. Over $46 million a year are spent on tranquilizers in Britain.

ANTIDEPRESSANTS
Each year in the United States over 5 million prescriptions are made out for antidepressants. The side-effects of these drugs include drowsiness, nausea, tremors, and constipation.

ANTIHISTAMINES
These are commonly used to treat allergic reactions such as hay fever. Antihistamines are also an ingredient in many cold "cures". Their side-effects can include drowsiness and blurred vision.

COLD "CURES"
Although there is no cure for the common cold, more than $145 million are spent on cold medications every year in Britain. Often a simple painkiller and plenty of fluid works just as well.

Drug sales worldwide have risen astronomically and today the production and sale of drugs has become less a matter of combating disease, and more a matter of making money.

Vast amounts of money are spent on developing new drugs and promoting them with expensive publicity campaigns so that they secure a position in a lucrative market. Doctors are subjected to enormous pressure from the pharmaceutical companies. It is estimated that they spend $43,000 a year on each doctor in Britain in an effort to persuade these doctors to test out, patronize, and publicize their particular brands. In the United States, advertising campaigns are not just directed at doctors through medical journals and the mailbox. The industry also uses television commercials to advertise their wares to practitioners.

But it is not only doctors who have come to put their faith in chemical medicine. Patients have as well. Far too many people now believe that chemical medicine is the only kind of medicine that can produce results.

THE SIDE-EFFECTS OF DRUGS

Chemical medicine is a powerful weapon that needs to be handled with care. Taking drugs to fight minor illnesses bypasses or even suppresses the body's own defense mechanisms, the mechanisms that should be capable of helping the healing process unaided. Many drugs are not even meant to fight disease; instead they are designed simply to overcome discomfort.

As with all synthetic chemicals, drugs bring unwanted side-effects. These range from the inconvenient to the tragic. Thalidomide, to cite the most devastating example, alleviated symptoms of morning sickness in pregnant women. For centuries this had been quite adequately treated with herbal remedies such as peppermint, camomile tea, or meadowsweet, but Thalidomide claimed to do the task more efficiently and moreover it was *modern* and it was *advertised*. So on the advice of their doctors, many thousands of women took the new drug. Their deformed children, now into adulthood, bear the burden of a drug company's appalling mistake.

So far the Thalidomide tragedy has not been succeeded by an event of similar proportions. But the side-effects of popular drugs remain a grave cause for concern. When they first appeared in the forties, steroids were hailed as a miracle drug because of the dramatic improvement they had on asthmatics and arthritics. But the delight faded as it became clear that the extreme discomfort caused by the illness returned as badly as ever once the effect of the steroids wore off. Furthermore, it was necessary not just to repeat the prescription, but to give patients increasingly strong doses to create the same effect. In other words patients became dependent.

With this growing dependency on steroids came a whole range of side-effects: skin rashes, excess body hair, bruising, bone and muscle decay, rising blood pressure, swelling of the face and trunk, and even heart problems. But the failure of one drug leads to the creation of another. In the United States, where doctors and drug companies are frequently sued by patients who have suffered from the side-effects of drugs, the American Medical Association is increasingly encouraging the use of alternatives to steroids. Nonsteroidal anti-inflammatory drugs are now frequently prescribed for arthritics instead of steroids, but they too are by no means harmless. One of them, Phenylbutazone, which was prescribed mainly for arthritis, caused the deaths of 440 people in Britain before it was withdrawn. Yet drugs of a very similar nature are still being prescribed in Britain and other countries.

COMPLEMENTARY MEDICINE

The medical profession, like many other walks of life, is in the hands of the "specialist." A specialist is a person who knows a tremendous amount about one thing, but often at the expense of knowledge of the whole. Specialists view the body as a collection of parts, each of which can be treated in isolation with drugs or surgery.

We are fortunate to be living in a time of reaction against all this. Complementary or holistic medicine, call it what you will, takes the opposite stance, and takes an interest in the patient as a whole. There has been an explosion in new ways of promoting health, some good, others eccentric. But the important thing is that it is happening.

The struggle between the medical establishment

and unorthodox ideas has been going on for hundreds of years. In 1630 local Amerindians first showed the bark of the Peruvian cinchona tree to the Spanish Jesuit, Father Calanch. It proved to be a most effective treatment for malaria which was rife at the time in Madrid, Lisbon, Rome, London, and Paris. But its success was not greeted with enthusiasm by the medical profession which did very well out of rather less effective "treatments"

for the disease. A British practitioner, Thomas Norton, has described it thus: "Opposition to Peruvian bark was mainly a result of a conspiracy between physicians and apothecaries who resented the cure of a disease which had for so long been an unmixed financial blessing." This statement has a familiar ring about it. The modern medical establishment too can be rather resentful of simple treatments for disease.

In the United States, research into a possible connection between food allergies and behavioral disturbances in children was undermined by weighty scepticism from orthodox practitioners who, with the moral and financial backing of the pharmaceutical companies, preferred to stick to conventional drug treatments for these patients. Although it costs medical authorities a lot of money to continue using expensive drugs, it greatly benefits the pharmaceutical companies, and after all it is they who put up much of the money to enable medical research.

Faced with the miserable effects of orthodox treatments for diseases such as cancer and arthritis, many people are turning to alternative forms of therapy that do not use these powerful chemicals.

While conventional treatment is mostly about the administration of drugs that suppress and relieve symptoms, practitioners of holistic medicine are mostly concerned with getting to the root of the problem – with eradicating the causes, not merely alleviating symptoms.

TREATING THE BODY AS A WHOLE

The aim of all alternative therapies is to boost the body's natural resources so that it can heal itself. This might mean releasing tension through massage or reflexology, relieving pain with acupuncture, eliminating accumulated toxins in the body by fasting, naturopathy, herbal treatments, or macrobiotics.

Alternative medicine certainly works. After years of painstaking research, Dr. Max Gerson came to the conclusion that chronic degenerative diseases were largely the result of bad nutrition and were often caused by eating inferior food grown on soils depleted of humus and trace minerals. He found that health could only be restored by proper

---

CURES WITHOUT CHEMICALS

This table explains the basic principles behind five alternative therapies which have become increasingly popular in recent years. None of them involves the use of potentially harmful chemicals.

HERBALISM

Herbal remedies are derived from the seeds, stems, leaves, and extracted juices of plants. Blended specifically for each patient, they create less risk of allergic reactions and side-effects than synthetic drugs.

YOGA

Yoga seeks to establish a balance between mind and body. It teaches relaxation and has great value as a preventative medicine, since many illnesses are caused by stress and anxiety.

REFLEXOLOGY

Reflexology holds that points on the feet are physically related to other parts of the body, both internal and external. By massaging the relevant point on the foot, problems elsewhere in the body can be alleviated.

ACUPUNCTURE

Acupuncture works on the principle of a balance between a positive and a negative force in the body. Imbalances cause emotional and physical ill-health, and are corrected by piercing points of the body with needles.

OSTEOPATHY

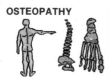

Osteopathists believe that the structure of the body must be correct before it can function properly. The positioning and movement of the bones is corrected by manipulating the joints.

nutrition and nontoxic medication.

Dr. Gerson devised therapies largely based on a pure, natural diet to restore the sick person's body as a whole, not just to treat a particular symptom. Its purpose was to enable the body to heal itself, not to "fix" it by stuffing it full of pills. Gerson had considerable success in the dietary treatment of cancer. His work however was largely ignored and even ridiculed by the medical authorities in the United States, where he spent the last 20 years of his life. Instead, the major research efforts were directed at radiation treatment, surgery, and chemotherapy. It is only much more recently that a bad diet has been recognized as a causative factor in the emergence of cancer by U.S. research organizations.

Dr. Bircher-Benner, who we have encountered in chapter 4, implemented similar theories at his clinic in Switzerland where he treated a range of degenerative illnesses with a wholefood diet based on grains and raw fruit with considerable success. The Bristol Centre, a cancer clinic in Britain, also bases its treatment on the power of the body to heal itself once unburdened by the toxins that have accumulated as a result of a typical modern lifestyle. These toxins can disable the body in disastrous ways, because we are not what we eat, but what we actually assimilate from our food. By damaging the natural bacterial flora of the digestive tract, toxins leave the body more vulnerable to infections such as *Candida albicans*, with the result that nutrients are not absorbed as they should be. Medical approaches which attempt to rid the body of the toxins have had results when orthodox practitioners had lost hope.

HERBS AND HEALTH

The active ingredients of about 70 percent of the drugs used by conventional medicine are based on chemicals found in herbs. Many of these have been known and used for thousands of years. As long ago as 2500 B.C. a Chinese herbal listed more than 300 herbal remedies, while Gerard's *Herbal*, published in Europe in 1636, lists a total of nearly 4,000.

Today all the known medically active, plant-derived prescription drugs come from just forty species of flowering plants. Of the quarter of a million species of flowering plants known to exist, only about 5,000 – a mere 2 percent – have been thoroughly examined as a source of natural drugs. The modern drug industry owes an enormous debt to nature for this source of raw material. Despite endless attempts it has not yet managed to improve on nature's recipes.

Only seven major pharmaceutical drugs can be synthesized more cheaply than they can be extracted and purified from plants. In Britain over one-quarter of all medicines used contain one or more active components derived from tropical plants and in West Germany they are present in half of the medicines in use.

Because they are rooted in the ground and cannot simply run away, plants have developed complex ways to defend themselves. They have had millions of years to evolve a huge variety of chemical defenses and it is these substances that give plants their curative properties.

Much of the knowledge of herbal remedies originates not from specialists but from popular usage. Throughout the ages there have been herbal practitioners who have experimented with medicinal plants from their local environment.

POSITIVE ACTION
How to keep healthy in an unhealthy environment

- **Eat healthily**
  Make sure that your diet is varied and contains enough vegetables and unrefined foods. Good nutrition improves the body's ability to cope with infections and stress-related disorders.

- **Exercise regularly**
  Regular hard exercise counteracts the bad effects of a sedentary life. It helps to fight stress-related problems by reducing physical and mental tension. Also, by increasing the body's metabolic rate, it may help to speed up the elimination of environmental toxins.

- **Avoid polluted air outside**
  Don't expose yourself to outdoor air pollution unnecessarily. Avoid exercising near sources of air pollution such as busy roads – you may do yourself more harm than good.

- **Avoid polluted air indoors**
  Polluted indoor air can have a bad effect on health, so avoid household chemicals (see chapter 8).

A natural cure
In China traditional and modern medicines are used in conjunction rather than in opposition to one another. Many hospitals, such as this one in Soochow, make extensive use of medicines derived from plants. The ingredients are carefully weighed and mixed to prepare remedies that contain dozens of active ingredients, unlike the single-substance remedies used in Western medicine. Herbal medicine brings a low risk of side-effects or addiction.

Herbal remedies have proved their efficacy and safety in common use over the centuries. No modern drug that is put on the market can be claimed to have been tested *that* thoroughly.

Some tropical herbs can be used to treat new diseases – such as heart disease, some viral infections, and even cancer. Lymphocytic leukemia, for example, is now being treated very successfully with a drug derived from the rosy periwinkle, a native of the West Indian rain forests. Sufferers of this disease, which was fatal until recently, are now almost certain to recover. Chemicals derived from the rosy periwinkle have also been used successfully in treating several types of cancer, including Hodgkin's disease.

So what is being done to make the most of this treasure-house of natural remedies? Very little. Instead, their habitat is being cleared and plant species are dwindling. This is a reckless waste of resources. According to Friends of the Earth, at least 1,400 plant species from tropical forests are believed to offer potential for cures against various cancers. How many more will be lost before we have a chance to recognize their virtues?

THE FUTURE FOR PLANT-BASED MEDICINE

Why don't drug companies use natural products rather than those made by synthetic processes?

Well, apart from the argument that some synthetic drugs are cheaper, there is another very important financial explanation for the continual pursuit of new drugs: not patients, but patents.

Despite the vast number of plants awaiting investigation, many of which come from the increasingly endangered tropical forests, the medical establishment and the drug companies have little interest in proving their usefulness. They prefer to concoct "new" substances because these can be patented. It is not nearly so easy to patent a substance that Mother Nature has been producing for millions of years already.

Herbs contain combinations of powerful chemical substances, the uses of which have been demonstrated by generations of skilled herbalists, although often not "proven" by modern scientific techniques. *These* are what the drug companies should be investigating instead of turning out yet another brand of headache remedy to add to the thousands we already have. Plant medicine is our heritage. We should protect the places where these plants grow and use them in preference to synthetic drugs wherever possible. Real health does not come out of chemical factories.

8

THE HAZARD-FREE HOME

Unseen hazards about the home
Cleaning up the indoor air
How to reduce paint pollution
The plastic-free house
The benefits of natural materials
Safeguarding wood for the future

Until concrete and plastic appeared on our planet, houses were built entirely out of natural materials. They were products of the earth and to the earth they often returned. The "ingredients" that made up the ordinary person's house were those that could be obtained locally and with the least trouble. The walls were made of timber, stone, rammed earth, or bricks. Wooden planking, tiles, stone slabs, or clay made up the floor, while the roof was constructed out of thatch, wooden shingles, tiles, or some kind of worked stone. Furthermore, everything inside the house was made of natural materials too. The furniture was made of wood, covered and padded with wool, linen, and cotton, while the carpets and matting, if

any, consisted entirely of what could be produced from plants and animals.

Now while the occupants of such houses did not always enjoy the comfort that we do today, they did have some advantages over us. And that was a complete understanding of the materials that made up their homes and a knowledge that all had stood the test of time and were perfectly safe.

The houses of today bear little relation to their traditional counterparts. Our homes are made of, and contain, substances that have only recently been invented, and the full effects of which we are only just beginning to appreciate. Houses have been transformed by synthetic materials. A large proportion of the plastics produced by the chemical industry end up in the home as tabletops, foam mattresses, cushions, furniture, fabrics, wallpaper, and so on, while in decorating we use many chemicals that were unknown a few decades ago. Just what side-effects all these will have, no one really knows.

BUILT-IN HAZARDS

In the past, the safety of a house was judged simply by whether it stood up or not. As long as it stood, all was well. But since then things have become more complicated because even the best-built house can be unsafe or unhealthy by reason of the products that go into it.

Many of the materials used in construction are hazardous to the world at large during their manufacture. Producing cement for use in concrete blights land by choking it with dust; baking bricks can create poisonous fumes which contaminate vegetation and in turn the milk of cows. There is not a great deal that can be done to prevent this. However, once concrete, aggregate blocks, or bricks are in place in a house, there is little evidence that they do any lasting harm to the house's occupants.

Matters are rather different with some more recent inventions. When it first appeared, asbestos was hailed as a perfect building material. It was nonflammable, light, and easily cut into shape. Asbestos roofing sheets and asbestos paneling appeared the world over. But then the medical profession realized that asbestos fibers were a major health hazard, causing lung cancer and other diseases in production workers and in anyone regularly in contact with the material.

In 1982, the largest asbestos producer in the world, Johns Manville Corporation, went bankrupt. The company declared itself unable to pay damages to former employees and customers. Johns Manville had paid out nearly $50 million to over 3,500 claimants; another sixteen thousand cases were left unsettled. These people were just a fraction of those at risk. It has been estimated that in the next 20 years between 160,000 and 200,000 workers will develop asbestos-related cancer.

Now when asbestos was first produced, no one could possibly have predicted that it would have

AS SAFE AS HOUSES:
The environmental problems of the modern home

Most people spend two-thirds of their lives indoors, half of this being at home. For this reason, the home is a very important environment: what goes into building a house and fitting it out has a direct effect on the health of its occupants. It also has important consequences, through pollution and the use of raw materials, for the world beyond the front door.

Imported timber

HABITAT DESTRUCTION
No one can possibly build a house without harming natural habitats – but in too many cases, land is ravaged quite unnecessarily with no attempt being made to build around what is already there.

THE RAIN FOREST CONNECTION
Much of the hardwood timber used in house construction comes from the tropical rain forests of Southeast Asia. Using tropical hardwoods further endangers an already hard-pressed habitat.

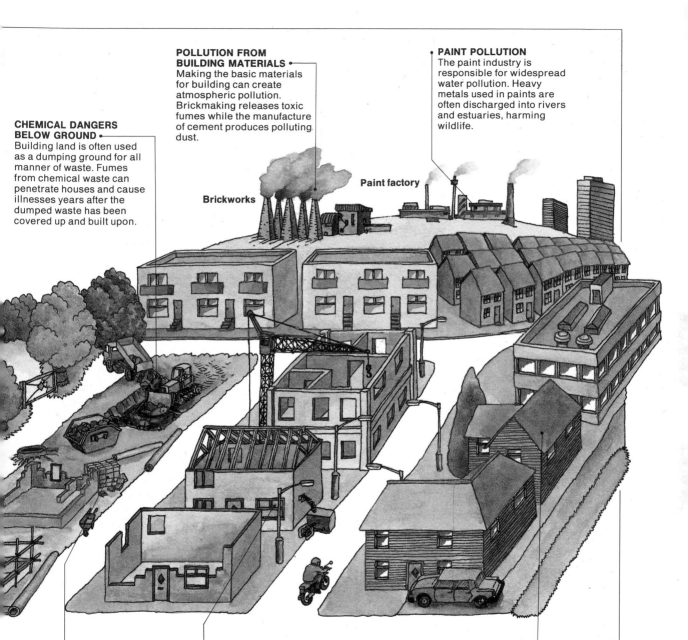

POLLUTION FROM BUILDING MATERIALS
Making the basic materials for building can create atmospheric pollution. Brickmaking releases toxic fumes while the manufacture of cement produces polluting dust.

PAINT POLLUTION
The paint industry is responsible for widespread water pollution. Heavy metals used in paints are often discharged into rivers and estuaries, harming wildlife.

CHEMICAL DANGERS BELOW GROUND
Building land is often used as a dumping ground for all manner of waste. Fumes from chemical waste can penetrate houses and cause illnesses years after the dumped waste has been covered up and built upon.

Brickworks

Paint factory

PLASTICS
Plastics make up an increasing proportion of the materials used in building. They cause outdoor atmospheric pollution during their production, and some of them, such as PVC, produce indoor atmospheric pollution once put into place.

DANGEROUS BUILDING MATERIALS
Materials such as asbestos and urea-formaldehyde insulation foam, have both proved hazardous to health after long-term exposure.

INDOOR AIR POLLUTION
The air inside the average house contains many substances not present in the atmosphere outside. It is often contaminated by fumes from solvents, paints, and other household chemicals.

CHEMICAL HAZARDS IN THE HOME

Decorating and maintenance bring a host of potentially lethal chemicals into the home. Some of these carry warnings advising of the dangers in their use, but many do not. Choosing the wrong product or applying a toxic substance carelessly can turn your home into a lived-in health hazard.

ADHESIVES
Petroleum-based adhesives produce a toxic vapor. Most other adhesives contain cyclic hydrocarbons – compounds which are frequently linked with cancer.

FUNGICIDAL PAINT
Long-acting fungicides are often used in paint intended for damp conditions. But in kitchens the fungicide may be transferred from walls, via the hands, to food.

BAD VENTILATION
Inadequate ventilation magnifies indoor pollution enormously. In these conditions the solvents used in home repair, which evaporate rapidly, build up to high levels.

INSULATION HAZARDS
Urea-formaldehyde foam, which is pumped into cavity walls as a liquid which then solidifies, has been shown to give off poisonous formaldehyde gas. This may seep through walls and reach high concentrations in a badly ventilated room.

PAINT SOLVENTS
All paint solvents are highly dangerous if inhaled. In addition, their production brings with it the risk of dermatitis and, in severe cases, brain damage.

WASTE SOLVENTS
When poured down the drain, substances such as turpentine and paint stripper are carried to sewage treatment works. There they poison the bacteria that decompose waterborne waste.

BONDING AGENTS
Much of the wood used in homes is made up of wood chips which are bonded together by a formaldehyde-containing resin. This resin produces fumes for many years after the composite wood is installed.

DANGEROUS DUST
Using any form of abrasive can liberate dust into the atmosphere. This may contain poisonous wood preservatives, poisonous metals from paint, and particles of bonding agents used in wood composites.

PAINT POLLUTION
Although modern paints are not as as their predecessors, their manufa often involves pollution of rivers an coasts by poisonous effluent.

POSITIVE ACTION

Avoiding chemical hazards in the home

- **Use safe insulation materials**
  Glass fiber is an effective and nontoxic alternative to plastic insulation (polyurethane foam or expanded polystyrene). Its production is relatively pollution-free.

- **Take care with adhesives**
  Many quick-drying adhesives not only produce vapor which is damaging to the lungs; substances in these glues may also be absorbed directly through the skin. White latex-based glues are much safer than those which contain volatile solvents.

- **Dispose of household chemicals safely**
  Take care when throwing away leftovers. Some local authorities operate special facilities for the disposal of household chemical waste. If yours does not, make sure that all paint and solvents are well wrapped before putting them in the trash can.

- **Use low-toxicity wood preservatives**
  Heat treatment can protect structural timber from fungi and insects, and may be available from your timber specialist. Otherwise, specify a low-toxicity preservative. These are usually guaranteed for at least 10 years.

- **Avoid fungicides**
  Fungal attack is caused by damp. Using a fungicide only eradicates the symptoms—it does not cure the problem. A safer investment is adequate ventilation or damp-proofing.

- **Be safe with solvents**
  As a general rule, the less you use of these the safer your home will be. Do not buy solvents of any kind in excessive quantities.

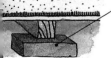

WOOD PRESERVATIVES
These are long-acting poisons. Treated wood is poisonous if ingested, while wood preservative applied by spraying may contaminate nearby objects.

HAZARDS FROM SYNTHETIC MATERIALS
Today's home contains large numbers of synthetic materials, like the man-made fibers of carpets. These may produce hazardous gases when brought into contact with household chemicals like paint solvents.

such disastrous results. Cause and effect were too far divided by the passage of time. The asbestos disaster should have been a signal for extreme caution when using new materials in houses, or any other buildings for that matter. But the signal has gone unheeded. New products appear every year, and many people are quite probably being harmed at this very moment by their hidden effects.

POISONS OF THE INDOOR AIR

The dangers of synthetic materials in the home are never more apparent than when a fire breaks out. Today more people are killed by toxic fumes in house fires than succumb to the direct effects of fire itself. (This is one reason why firemen wear breathing masks as a matter of course when tackling blazes.) But your home doesn't have to catch fire for you to be affected by fumes from synthetic products. Indoor air is polluted by them all the time. And of these air poisoners, one has become especially notorious.

Many household materials contain formaldehyde. It is found in insulating foams, synthetic resins, disinfectants, medicines, glues in plywood, chipboard and hardboard, laundry detergents, and even some cosmetics. Pure formaldehyde is an extremely unpleasant gas which when dissolved in water gives off a pungent vapor. In the United States, its largest consumer, 3 million tons are produced every year.

This substance is now recognized, somewhat belatedly, as a dangerous pollutant. It has in one way or another injured a great many people. In 1982, a founder member of the U.S. Formaldehyde Victims' Association spoke to a special commission of the House of Representatives about its effects. She explained how her whole family had suffered from chronic colds, tiredness, and throat inflammation. Her husband was affected by a crippling bout of arthritis. Her children were unable to concentrate and to absorb what they were taught at school. All this was the result of using chipboard panels containing formaldehyde for dividing walls.

Formaldehyde is released very slowly. It seeps out of chipboard, insulation foam in cavity walls, and even furniture. Medical research at Harvard University has shown that regular exposure to concentrations of less than one part per million can cause defects of the nervous system and severe memory loss in adults.

By 1980 the U.S. Chemical Institute of Technology had established beyond reasonable doubt that inhalation of formaldehyde fumes caused cancer in rats. But because no *humans* had been shown to develop cancer from this form of indoor air pollution, there was no immediate action.

Canada and some American states have now outlawed insulation foams containing formaldehyde. In West Germany chipboard containing formaldehyde-based glue has been banned in public buildings. But in most countries there are virtually no restrictions on the use of materials containing formaldehyde. On the contrary. Advertisers and store owners will actively encourage you to buy products containing this substance.

ENDING FORMALDEHYDE POLLUTION

There is no need for anyone to be exposed to formaldehyde pollution. It is perfectly possible to make chipboard that does not contain this insidious substance. There are also a number of ways of insulating a house without using foam that produces formaldehyde. Mineral wool and glass fiber will work just as well without causing any air pollution. Formaldehyde-free cosmetics and laundry detergents do exist, although without a list of chemical ingredients for all the products in a store you may find it hard to pick them out.

Why is formaldehyde still used in such quantities? Because the chemical industry has so much at stake in it. For this reason alone it may be some time before formaldehyde is used only for its original purpose: preserving corpses.

SOLVENTS IN THE HOME

Formaldehyde is not the only artificial substance to produce dangerous fumes indoors. Household air is also contaminated by other products. Excluding tobacco smoke, a hazard easily dealt with, probably the most dangerous of these are the solvents.

Wherever paint is used, so are paint removers, paint strippers, brush restorers, and a host of other solvents. Most of us have a bottle or two of these substances tucked away on a shelf: added together, there are millions of gallons in homes worldwide.

REDUCING INDOOR AIR POLLUTANTS

This chart identifies eight major sources of air pollution in the home. In each case only the main pollutants are detailed; some sources, like cigarettes, produce dozens of different airborne substances which all pollute the indoor air.

Cigarettes These produce a range of poisonous gases – carbon monoxide, ammonia, nitrogen oxides, and complex organic chemicals.
Action Stop smoking!

Gas stoves These stoves can produce carbon monoxide and nitrogen dioxide – both of which can cause respiratory problems.
Action Ensure that ventilation is adequate when stove is in use.

Electrical equipment High-voltage equipment and household appliances with bad connections can produce toxic ozone.
Action Make sure that all appliances are securely wired.

Chipboard The resin used as a bonding agent releases hazardous formaldehyde gas.
Action Use solid wood wherever possible; ventilate rooms containing new chipboard.

Aerosols These contain chlorofluorocarbon propellants. These are an indirect health hazard through their effect on the ozone layer.
Action Avoid using aerosols.

Solvents Paint thinners and strippers contain poisonous hydrocarbon solvents. These are highly persistent in indoor air.
Action Always use adequate ventilation.

Open fires Inadequate air circulation can allow the products of partial combustion, including carbon monoxide, to accumulate.
Action Keep flues clean, so that all fumes are carried outside.

Insecticides Many insecticides contain toxic organic chemicals which can accumulate in the body.
Action Avoid insecticides; remove the food traces that attract pests.

Solvents: the two-fold addiction
Solvents have become drugs of easy access. Because so many household products – from glues to typewriter correction fluid – contain solvents, ensuring that none of them are misused is very difficult. The fact is that the average householder is a solvent addict – albeit of a rather different kind. Huge quantities of solvents are stockpiled in almost every home.

Most solvents are hydrocarbons or chlorinated hydrocarbons (members of the same family of chemicals that contains many pesticides). When they are used, they evaporate rapidly and, try as you may, it is hard to avoid breathing them in.

Just how safe these solvents are can be deduced from the history of the dry cleaning industry. Clothes are dry cleaned by immersing them in a solvent. When the first dry cleaners opened in 1845 in France, benzene was used as the cleaning fluid. Besides having an annoying tendency to catch fire, this also gave dry cleaning workers leukemia. So different solvents were tried. Carbon tetrachloride and tetrachloroethane were next, but again these proved fatal. A succession of solvents came and went. Today, one of the most popular is PER, or perchloroethylene. This may not be a hazard to the people who take the occasional garment to be cleaned, but long-term exposure to it leads to nervous disorders and damage to the lungs and kidneys. It also harms trees and other plants.

This cautionary tale throws a long shadow over a host of household products that either consist of, or that are dissolved in, chlorinated hydrocarbons. Household glues can contain them. The epidemic of glue sniffing which has engulfed many countries in recent years shows all too vividly what they can do to the body. When solvents are accidentally spilled on other man-made products, like carpets made of synthetic fibers, even more chemicals are likely to escape into the air.

Solvents not only pollute the air you breathe – if they are poured down drains or on the ground they also find their way into rivers and aquifers, and from there into drinking water. As with so many toxic chemicals, the remedy is to use as little of them as possible.

WOOD PRESERVATIVES

Wood preservatives contain potent fungicides and insecticides. Across the world, about 100 million square feet of timber surfaces are treated with them every year.

Many wood preservatives contain pentachlorphenol (PCP) to prevent and kill fungi and lindane, a potent insecticide. PCP is a very powerful poison, indeed, it is designed to be. Applied to timber it is meant to keep fungi at bay by killing the spores. It remains active for many years and is a health hazard not only to people in their homes, but also to the workers who make the chemical. Cirrhosis of the liver, bone marrow atrophy, and nervous disorders are all part of the price paid to produce this common household poison.

Although lindane is restricted as an agricultural insecticide, many countries still allow its use in the

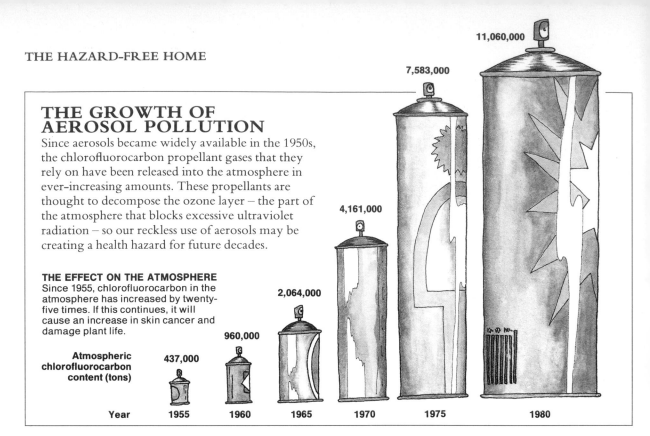

THE GROWTH OF AEROSOL POLLUTION

Since aerosols became widely available in the 1950s, the chlorofluorocarbon propellant gases that they rely on have been released into the atmosphere in ever-increasing amounts. These propellants are thought to decompose the ozone layer – the part of the atmosphere that blocks excessive ultraviolet radiation – so our reckless use of aerosols may be creating a health hazard for future decades.

THE EFFECT ON THE ATMOSPHERE
Since 1955, chlorofluorocarbon in the atmosphere has increased by twenty-five times. If this continues, it will cause an increase in skin cancer and damage plant life.

Atmospheric chlorofluorocarbon content (tons)

Year	1955	1960	1965	1970	1975	1980
	437,000	960,000	2,064,000	4,161,000	7,583,000	11,060,000

home. Sweden, Holland, Japan, and some U.S. states are exceptions: here it has been banned altogether. People living in these more enlightened places not only escape lindane poisoning at home they also escape being poisoned by its by-products.

Chemical companies dispose of these by-products with varying degrees of success. In West Germany there was a public outcry when 100,000 tons of toxic residues from lindane production were discovered in the grounds of the Merck chemical company in Darmstadt. Some of it had been stored on an open dump and from there the dust had blown on to neighboring farmland, heavily contaminating it. In Hamburg the chemical company Boehringer stored lindane residues in the works' grounds and on the nearby and notorious garbage dump at Georgswerder. For years the factory released the poison into the air from its smoke stacks. Vegetables grown near the factory were tested in 1979 and declared to be unfit for human consumption and were destroyed.

In 1984 the Boehringer factory was closed down as a serious danger to human health. Among other things its chimneys emitted the dioxin TCDD – the substance we have already encountered in the problem of waste disposal. This dioxin is an impurity that comes of making the two main ingredients of wood preservatives, so anyone who buys preservative that contains PCP or lindane simply adds to this poisonous waste.

ALTERNATIVES TO POISONOUS PRESERVATIVES

Once you have coated your home with poisonous wood preservatives, there is no undoing it. But is it actually necessary to use preservatives in the first place? After all, Europe has many timber-framed houses that have survived since medieval times without a drop of preservative. Why do we suddenly need them now?

Well, under most circumstances we do not. As long as wood is kept dry, neither fungi nor insects are likely to become major problems. It is possible to protect timber from moisture by applying varnish, lacquer, or beeswax. Timber can also be treated with preparations containing borax, acetic acid, or soda, all of which are safe to use indoors.

If you do live in a house that has decaying timber, there is no need to reach for the most poisonous chemical available. There is far too much "overkill" in the timber treatment business. A low-toxicity chemical that is guaranteed to remain effective for ten years will do the job quite as well as one that lasts two or three times longer, but much more safely to those inhabitants of the house who are neither insects nor fungi.

You may be able to do without chemicals altogether. In some countries hot air treatments are available to combat woodworm and fungi. Hot air is circulated in a room or loft for a few hours by a special hot air blower. Insects and fungal spores are killed without any dangerous fumes remaining to contaminate the house. Furniture can be treated quite simply by being put in a small heated space like a sauna or by being soaked in a bath of hot liquid wax.

PAINT POLLUTION

At one time lead was a major constituent of paint. Indeed, if you live in a house over fifty years old, it is likely that the walls have had a good quantity of lead slapped over them by its previous occupants. But if you think that problems with paint are entirely confined to the past, you are wrong. Paint still presents health risks, and it is still one of the most polluting products used in the home.

Lead used to be added to paint first and foremost to help it to dry, although sometimes it was used to add color or prevent corrosion. But there was actually no need for any of this lead to be used. The amount of lead in different brands used to vary enormously with little noticeable effect on the quality. Despite much clamor on the part of the paint industry, lead levels were eventually restricted and today low-lead household paints are the only kind permitted in the United States and most European countries.

Because paints containing lead were used for so long, a lot of lead still surrounds us on painted walls. As these are stripped down and repainted, the lead finds its way into the air, drains, and into the soil. All old paint needs to be treated with caution, whether still in a tin, on walls, or on woodwork.

Unfortunately paint technologists have managed to come up with some more poisonous metals to go in paint. One of these is cadmium. Thousands of tons of it are used every year, and a large proportion

---

POSITIVE ACTION
How to minimize painting hazards

- **Check that your paint is lead-free**
  All household paints made today are lead-free: old leftover paint may not be. Before using old paint, check its lead level with the manufacturer (the maximum safe level is about 600 parts per million). Emulsion paints do not contain lead.

- **Precautions when sanding**
  Old paint may have a high lead content. Wear a mask when sanding down old paint and ensure that no paint flakes are left where children may find them.

- **Paint strippers**
  Paint strippers eliminate the hazard of airborne lead particles from old paint but give off toxic vapor. Ensure that a room is well ventilated before using them. Wrap up all scrapings before throwing them in the trash can.

---

of this ends up on walls inside homes. Titanium dioxide, too, is used as a pigment in household paints. Ironically this was introduced as a result of the growing unease about lead in paint. Titanium dioxide is a white pigment which is used in synthetic oil paints and emulsions. In the production of paints containing titanium dioxide, liquid wastes containing sulfuric acid, heavy metals, and chlorinated hydrocarbons are produced and have to be gotten rid of. In the last few decades the toxic leftovers of the paint industry have become a major environmental hazard. Hundreds of thousands of tons have been dumped into the North Sea from special tankers, or piped out into estuaries. The damage to fish and other wildlife has been profound.

GENTLE CHEMISTRY

Every year, about 20 million tons of organic (carbon-based) chemicals are produced in giant chemical plants. Up to 50,000 ingredients make up more than a million substances that circulate in these chemical factories, and hundreds of new substances are produced to fill alleged gaps in the market. Of the preparations destined for the home, each is apparently superior to all the ones that went before it.

Many household chemicals are highly reactive, highly dangerous substances. Their advertising emphasizes this by describing them as "extra strong," "powerful," "high speed," and "super concentrated." It is all made to sound very desirable.

But the very mention of strength or power, when applied to a household chemical, should make us all bristle with apprehension and caution. Chemicals made from dangerous ingredients should have no place in the home. For these are the products that time after time turn out to be long-term poisons or the agents of cancer.

What is needed instead are chemicals that are created from natural substances. There is no shortage of ingredients. Timber treatment liquids, paints, glues, and cleaning fluids can be made from plant substances such as turpentine, pitch, latex, beeswax, natural oils, and dyes. The processes by which these are made would avoid many of the dangers that attend chemical factories, with their huge supplies of toxic raw materials which are combined under extremes of temperature and pressure. The gentler chemistry of natural products could create equally useful substances with far less danger.

USING NATURAL MATERIALS

Because synthetic materials in the home bring with them so many problems, the time is surely right to turn back to what nature has to offer. There is no point in poisoning ourselves and cluttering up our planet with plastic if nature can supply much better raw materials in the first place.

What is needed is a new type of architecture – one that is concerned with providing more natural homes. Such a thing does exist. In Germany, Austria, and Scandinavia, "biological" architects have been building houses that use naturally available materials in buildings wherever possible.

Making houses out of natural materials does not mean returning to the drafty, cold, and damp homes of a hundred or more years ago. A natural house can be quite as comfortable as one made out of plastic and concrete, and it almost goes without saying that aesthetically it will be far superior.

NATURAL ARCHITECTURE

Natural architecture is a matter of making sensible choices. If you are building a house and wish to keep to natural materials, bricks are an acceptable alternative to concrete, as is timber if the supply is plentiful. The cavity between the inner and outer walls should certainly be insulated, but not with plastic foam. By far the best insulator is recycled paper or cork, but rock wool or glass fiber will also do. None of these, of course, will produce poisonous gases, and none will be left to pollute the planet if, in the distant future, the house is demolished.

If the house is to be built where rain is frequent, the roof should be pitched (the flat roof being a nonsensical invention for wet countries) and it should be covered with clay tiles, stone slabs, or slate. Underneath it, tarred paper should be used to form a second barrier against moisture.

The natural materials for floors are stone or wood. Over them, a layer of cork will provide excellent insulation. Another material that can be used for this is old-fashioned linoleum. This is not the plastic-based "lino" which clogs up garbage dumps, but a floor covering made from natural plant resins.

For every synthetic material used inside the home, there is a natural counterpart: this is hardly surprising because most synthetic products were designed as imitations of natural materials in the first place. Instead of carpets containing man-made fibers, they should be of wool, coconut fiber or sisal. The furniture should be made of wood that is joined in the traditional manner, rather than glued together. The wallpaper, if there is any, should be paper – not some paper and plastic combination which is impossible to dispose of. All the doors and cupboards should be made of wood, not wood chips or fibers glued together with a substance that will give off fumes for years and years.

Paint, where it is used, could be made from natural ingredients such as lime. All the indoor wooden surfaces could be coated with natural wax. The completed house would then be completely organic – made of nature rather than against it. If you lived in such a house there would be no possibility of it poisoning you or any future occupants.

THE PLASTIC-FREE HOME

Modern houses are full of plastics masquerading as natural materials – PVC boarding that pretends to be real wood, nylon carpets that look like wool, plastic wallcoverings that simulate paper. All this plastic creates pollution both inside (by polluting the air), and out of doors (during its manufacture). But hardly any plastic is really necessary in building. The natural materials that plastics are designed to imitate are usually just as effective, and environmentally far less of a hazard. With the possible exception of ground moisture barriers and electrical insulation, no house needs to have plastic ingredients.

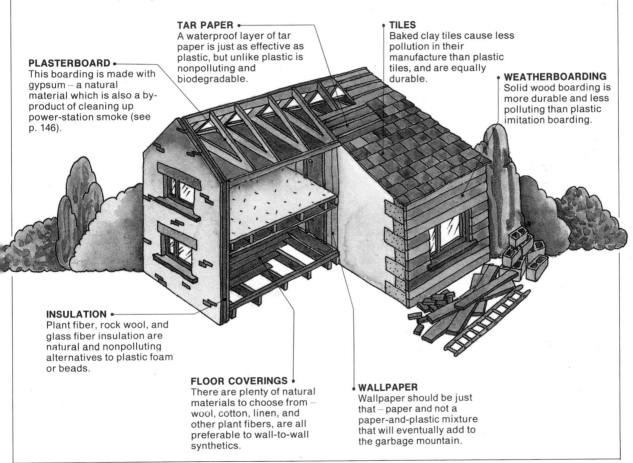

TAR PAPER
A waterproof layer of tar paper is just as effective as plastic, but unlike plastic is nonpolluting and biodegradable.

TILES
Baked clay tiles cause less pollution in their manufacture than plastic tiles, and are equally durable.

PLASTERBOARD
This boarding is made with gypsum – a natural material which is also a by-product of cleaning up power-station smoke (see p. 146).

WEATHERBOARDING
Solid wood boarding is more durable and less polluting than plastic imitation boarding.

INSULATION
Plant fiber, rock wool, and glass fiber insulation are natural and nonpolluting alternatives to plastic foam or beads.

FLOOR COVERINGS
There are plenty of natural materials to choose from – wool, cotton, linen, and other plant fibers, are all preferable to wall-to-wall synthetics.

WALLPAPER
Wallpaper should be just that – paper and not a paper-and-plastic mixture that will eventually add to the garbage mountain.

MAINTENANCE WITH NATURAL MATERIALS

Living in a "biological island" is, at the moment, only for those who can afford to have a natural home made from scratch or who have the good fortune to take over a house that has escaped being "improved" by twentieth century materials.

But you don't have to move to enjoy some of the benefits of a natural approach to housing. Every time you decorate your home, you will also be faced with making choices about materials. It is a simple business to avoid what may be polluting and to use natural materials instead.

In practical terms, this means saying goodbye, as far as is possible, to materials made of plastic: plastic shelving, plastic furniture, and plastic-covered board. It also means avoiding reconstituted wood, so chipboard, blockboard, and all other glued wood products do not appear in the home. The

Wood – the natural material
Solid wood creates no pollution problems when it is fashioned. It is extremely hard-wearing and durable, and when it is discarded it will decay naturally. The use of wood is currently experiencing something of a revival, as more people come to appreciate how vastly superior it is to plastic. But to meet the increasing demand, much more effort must be put into growing it on a sustainable scale.

same applies to furniture: look for furniture made of solid wood rather than glued wood mixtures.

Paints and varnishes you also need to select carefully to avoid products with a high metal content. Built-in fungicides, such as are found in wallpaper paste, should also be left on the shelf. Pure cellulose wallpaper paste, however, is perfectly acceptable, being a wholly natural product. By choosing natural materials as far as is possible, your home will become gradually safer as years go by.

THE ENVIRONMENTAL EFFECTS OF USING WOOD

It might occur to some people that house building and maintenance with natural materials seems to call for an awful lot of *wood*. It could be asked, with some justification, where all this wood might come from, as forests the world over are a threatened resource.

Wood supply is indeed a problem already. Even in houses of the plastic-and-concrete school of architecture, large quantities of wood are hidden under plasterboard, insulation board, wallpaper, roof tiles, or carpets. Millions of tons of softwood timber are currently used in countries whose forest stock has already been depleted by decades or centuries of forest clearance. In Europe nearly all softwood timber comes from Sweden, Finland, Norway, or perhaps even from Siberia or northern

POSITIVE ACTION
Natural alternatives to indoor plastics

● **Carpets**
Synthetic materials in carpets make up the greatest quantity of "fixtures and fittings" plastics in the home. Polyurethane-backed carpets in particular are known to produce dibutyl hydroxytoluene – an irritating vapor. If you have synthetic carpets, replace them with natural fiber carpets when they wear out.

● **Bedding**
A large proportion of bedding materials are made from acrylic and other synthetic fibers. Choosing bedding and mattresses made from natural fibers is often no more expensive, and the result is just as comfortable, less polluting, and more healthy.

● **Clothing**
There is no evidence that artificial clothing fibers are a health risk, but they certainly add to the general environmental problems of plastics. Buy natural fiber clothing whenever possible.

● **Food containers and wrappings**
With the exception of plastic film (see p. 67) there is no clear evidence that plastic food containers are dangerous. But natural materials like cardboard and greaseproof paper are biodegradable, and are therefore preferable.

Canada. In North America, softwood timber is nearly all home-grown while a proportion of hardwood is imported.

Much of this softwood timber comes from forest plantations of fast-growing conifers. Until a few centuries ago, firs and spruce trees occupied only northern lands and the higher slopes of mountains, but in Europe today they dominate even lowland landscapes. Conifers rule supreme because they grow quickly and have a fibrous wood which is good for construction purposes (and also, incidentally, they are a very suitable raw material for paper and cardboard).

Of the fast-growing trees, the fastest of them all, the spruce, is king. In central Europe over half of the forest trees are spruces. Spruce plantations may be dark and gloomy places, inhospitable to people and to wildlife, but foresters maintain that these are the only forests that are economical because they produce the kind of timber we want today.

There is no doubt that endless ranks of conifers can cause long-term damage to the land if they are grown in places to which they are not native. At one time, both Europe and North America had huge areas of mixed broadleaved woodland. When this is cut down and replaced by conifers, not only is an important resource lost, but the land itself is impoverished. Softwood plantations leave the soil lacking in humus and make it increasingly acid and in the long term this makes it difficult to grow any other kind of crop on the land.

TIMBER FROM THE TROPICS

It is not only softwood that ends up in houses. Europe, North America, and Japan between them get through millions of tons of hardwoods every year, and much of this is turned into doors, windowframes, and furniture. Nearly all this wood comes from tropical forests. Tropical forests are being cut down not to supply the needs of local people, but to make way for farmland and to supply timber to those countries that have already put their own trees to the axe.

The supply of tropical timber is finite. West Africa (with the exception of Cameroon and Zaire) is practically logged out. The European appetite for mahogany, made famous by the master furniture makers like Chippendale, has caused the destruction of huge areas of tropical forest in Africa. It has even been suggested by climatologists that the loss of moisture from Africa's air as a result of the forest clearance in West Africa has contributed to the southward push of the Sahara Desert.

The Philippines, Malaysia, and Indonesia are the world's major exporters of tropical hardwoods. Some of the most luxuriant forests in the world, sometimes containing up to 150 species of trees per acre, are now supplying the hardwoods which we use when we install windowframes for double glazing in our homes.

Extracting this timber is a highly destructive business. In order to gain access to dense forest, gashes have to be cut into it. Modern chainsaws do not take long to cut through trees 150 feet tall, but as they come crashing to the ground they pull down the neighboring trees to which they are connected by lianas.

In tropical forest, perhaps 10 percent of the trees are valuable timber species. The rest is left as the bulldozers drag the huge trunks of the selected few across the delicate forest soil, compacting it and damaging the roots as they crunch their way toward the timber collection yard. Loaded on trucks the trees end up in sawmills or are taken directly to harbors for their journey to timber-hungry countries.

SUSTAINABLE TIMBER

The way in which softwood and hardwood is exploited in temperate latitudes and tropics alike is both ecologically damaging and unsustainable. It simply cannot last. Timber needs to be a properly managed resource, one that is constantly replenished. The more we use, the more we should plant. In keeping with the principle of localism, the timber we use in our homes should be the timber that can be grown in that part of the world we happen to live in.

Proper forestry doesn't mean just growing one kind of wood at the expense of all others. It means planting trees that provide softwood and grow quickly, but also planting trees that take longer to mature and that provide hardwood.

Tropical hardwoods usually grow much faster

Destination overseas
The timber shipped abroad from the tropics comes from a rapidly diminishing stock of wood. In 1950, about 15 percent of the world's surface was covered by tropical forest. The trade in tropical timber, together with the felling of trees to make room for agriculture, could reduce tropical forest cover to about 7 percent by the year 2000.

than broadleaved trees in temperate countries. But this is no reason not to start growing far more hardwood trees in Europe and North America. The fact that we have too much grain on our hands and too little timber shows that we are growing the wrong crops. For timber is a crop just like any other.

In many countries, there are subsidies available for planting broadleaved trees for hardwood timber. But so far more trees are being felled than planted. The world stock of hardwood trees should be built up, rather than depleted. It may well be that some money from the public purse is needed to establish more varied forests once again. But as taxpayers we should remember that forests are more than just places for producing timber. They protect the soil from erosion, particularly on steep slopes unsuitable for farming. They provide us with recreation, they filter the air, and they are a home for the wildlife which can no longer exist on our bare agricultural landscapes.

KEEPING WOOD HOME-GROWN

If more hardwood trees are grown in temperate regions, the demand for tropical timber will eventually fall. But what can be done to conserve tropical forests in the short term, and with them some two-thirds of the world's plant and animal species?

Well, we can avoid buying tropical timber products, first of all. A mahogany toilet seat is, after all, not essential for anyone's well-being. Softwood doors and windowframes should serve just as well as those made from tropical timber.

But tropical countries do require some foreign exchange and timber exports are one way of generating income. As long as the wood is for sale, it will be bought. So what can be done to make timber exports from the tropics less environmentally damaging?

The answer, as with our own forests, is careful management and replanting. Not replanting with tropical conifers and eucalyptus, but replanting with the trees that were there in the first place. Timber extraction from virgin tropical forests cannot go on forever, but sustainable forest plantations could yield timber almost indefinitely.

Starting a hardwood plantation is a slow business, but in several countries, otherwise unworked land has been set aside for new trees. This work will be helped enormously if plantation timber is bought in preference to the timber of forest trees. In Britain, Friends of the Earth and the Tropical Timber Trades Federation have reached a tentative agreement about a code of conduct on timber extraction and processing. This will make it clear to timber buyers which wood has been grown in a sustainable way. If everyone chooses this kind of tropical timber in preference to wild timber, the decline of tropical forests may be slowed down and perhaps arrested altogether. One of the greatest environmental hazards − the destruction of forests − may thereby be averted.

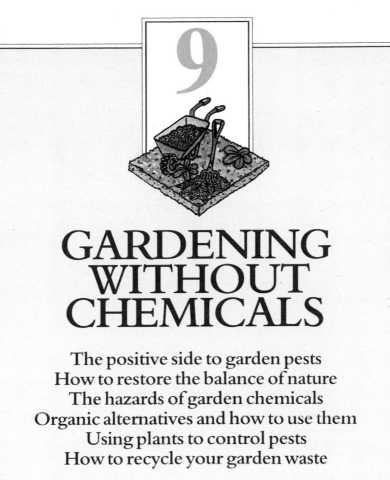

9

GARDENING WITHOUT CHEMICALS

The positive side to garden pests
How to restore the balance of nature
The hazards of garden chemicals
Organic alternatives and how to use them
Using plants to control pests
How to recycle your garden waste

A garden is a microcosm of the environment as we would like it to be. A beautiful and productive garden is the perfect example of a fruitful partnership between human intelligence and the rest of nature. It is an environment that we can partly control, and if we exercise that control in a humane way – for the benefit of other forms of life besides our own – we will find that we will more nearly approach the garden of our desires. But if we act like tyrants to other forms of life we deserve the fate of tyrants, which is to live and die unhappy, and alone. If we poison our environment we will ultimately poison ourselves.

Many a sundial tells us that "you are nearer to God in a garden than anywhere else on Earth."

Certainly for Mrs. Miniver, the rose-growing heroine of the American film made in 1942 that bears her name, love of gardening was what saw her through the constant crises and traumas of wartime Britain. But let us for a moment imagine a modern-day Mrs. Miniver in her floppy gardening hat, and her flower-patterned summer dress and dainty gardening gloves, tripping out to view her roses. How would she tend her garden now?

First she turns her attention to the weeds. She is going to use a herbicide containing a substance called trichlorophenoxyacetic acid, or 2,4,5-T. This is the stuff which mixed with 2,4-D made Agent Orange, which American forces sprayed over vast areas of jungle as a defoliant in the Vietnam war.

Thousands of children were born grossly deformed because of it. It also contains dioxin, which it seems is indispensable to its manufacture.

Undeterred, Mrs. Miniver gives the ground between her vegetables a good dousing with it. That should shrivel up the weeds. A half hour's gentle exercise with a hoe would have killed the weeds just as well – but that's so old-fashioned these days, isn't it?

Next she casts an eye over her roses. Aphids! Back she trots to the garden shed and emerges a very different Mrs. Miniver. Now her jaw is set like a rattrap and her eyes flash murder. In her hand she carries an aerosol spray can, inside which is a deadly poison, only known to man these last few years. After some half-hearted tests conducted by the company that manufactures and wishes to sell it, this poison has been "passed" for garden use. Like so many of its predecessors it will probably be banned in a few years, having been found to be carcinogenic. At Mrs. Miniver's heels trots her three-year-old son, but there are few worries for his health. For now, that is.

So Mrs. Miniver aims the spray at the offending aphids and blasts them. The merest whiff of the stuff would send them to eternity, but she drenches them. For good measure she soaks all the rose bushes, whether they have aphids or not. The fine spray drifts on the wind – on to her lettuces, on to her neighbor's lettuces, all over her little son. She is sure the stuff is safe – after all "they" wouldn't sell her anything dangerous, would they? As well as covering the aphids, it goes all over the ladybug larvae and the hover-flies and the *Anthocoris* bugs, all of which feed on aphids. Another generation of aphids will hatch and there will be no predators to control them.

If she had left the aphids alone she would still have gotten roses. If she had sprayed soft soap and water over her rose bushes, she would have killed that generation of aphids without harming the beneficial insects – nor her small son.

But Mrs. Miniver's blood lust is only partly assuaged and she goes into the shed again and comes out with a tin of chlordane. Many countries have banned chlordane because it is known to cause

THE GARDEN SUPPLY CHAIN
How gardening affects the environment

Modern gardening all too often is less a matter of producing and more a matter of consuming. Every year gardeners buy huge quantities of pesticides, soil conditioners, fertilizers, and plants. This cash-and-carry gardening can have damaging long-term results: it concentrates chemicals on small areas of land and encourages the removal of rare plants from the wild.

LOOTING FROM THE WILD
Endangered wild flowers such as Mediterranean bulbs and Central American cacti are plundered from their natural habitats to adorn gardens abroad.

STRIPPING THE SOIL
Organic soil conditioners such as peat and woodland humus take thousands of years to accumulate. Removing them destroys wildlife habitats which then cannot recover.

Mechanical extractor

Peat

cancer, but some have not. Mrs. Miniver soaks her lawn with it. She can't stand worm casts: they are so messy. No one has ever told her that these earthworms do only good, to her lawn or anywhere else where they are to be found. She is poisoning her best friends. A quick sweep round her lawn with a birch broom would spread the worm casts and turn them into a fine fertilizer. But she has a spiked roller and she makes her husband drag this over the lawn, in order to aerate it. The fact that the earthworms would have aerated it much more effectively, and for nothing, does not occur to her.

She has read the advertisements and paid careful attention to all the gardening programs. These never tire of telling her to solve all her problems in the garden with poisons – and if that doesn't work, with still more.

Next she turns and looks at her carrots. There is nothing wrong with them, but she heard last night on the radio of a creature called the carrot-fly. What if it struck? Remembering what the learned gentlemen said, she reaches for a proprietary substance containing diazinon. This is a persistent organophosphorus contact poison and more and more evidence is coming to light that it is a teratogen – an agent that raises the incidence of congenital malformations. Freaks, in other words. It is still legal in some countries, and she sprays the carrots generously with it. It is perhaps a little disturbing that Mrs. Miniver is pregnant.

WHY PREDATORS ARE IMPORTANT

The efforts exerted by such modern-day gardeners as Mrs. Miniver to destroy every single one of the pests and diseases in their gardens are completely unnecessary. There is simply no need to overreact the moment you see a greenfly or a thrip. In a healthy garden there is sure to be a predator nearby.

The orthodox approach to insects in the garden is: "By and large, insects are harmful to my interests. There may be some good ones, but I am not going to give them the benefit of the doubt. Anyway, I haven't got time to study them. Therefore I will get the strongest poisons I can."

Now nobody wants to see their cabbages ripped to pieces by caterpillars or their onions destroyed by

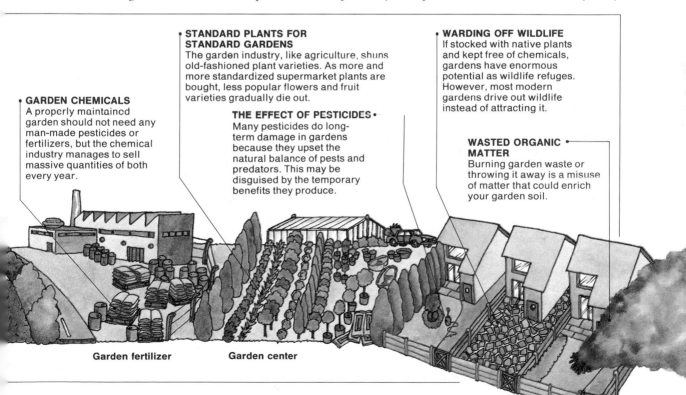

GARDEN CHEMICALS
A properly maintained garden should not need any man-made pesticides or fertilizers, but the chemical industry manages to sell massive quantities of both every year.

STANDARD PLANTS FOR STANDARD GARDENS
The garden industry, like agriculture, shuns old-fashioned plant varieties. As more and more standardized supermarket plants are bought, less popular flowers and fruit varieties gradually die out.

THE EFFECT OF PESTICIDES
Many pesticides do long-term damage in gardens because they upset the natural balance of pests and predators. This may be disguised by the temporary benefits they produce.

WARDING OFF WILDLIFE
If stocked with native plants and kept free of chemicals, gardens have enormous potential as wildlife refuges. However, most modern gardens drive out wildlife instead of attracting it.

WASTED ORGANIC MATTER
Burning garden waste or throwing it away is a misuse of matter that could enrich your garden soil.

Garden fertilizer   Garden center

THE CHEMICALS-AND-CONCRETE GARDEN

A garden is the householder's own private ecosystem to improve or destroy. In a well-ordered garden, nature and the needs of the gardener are carefully balanced. In a badly ordered garden, like the one shown here, nature is beaten into retreat in the senseless pursuit of neatness and tidiness.

HYBRID EVERGREENS
Modern fast-growing evergreens, with their tough leaves and inedible seeds, are almost completely useless to wildlife.

ORGANIC MATTER IN THE TRASH
Leaves, twigs, and uncooked food scraps are valuable organic waste from the garden and the home. They should be used for compost instead of being carted away at the taxpayer's expense.

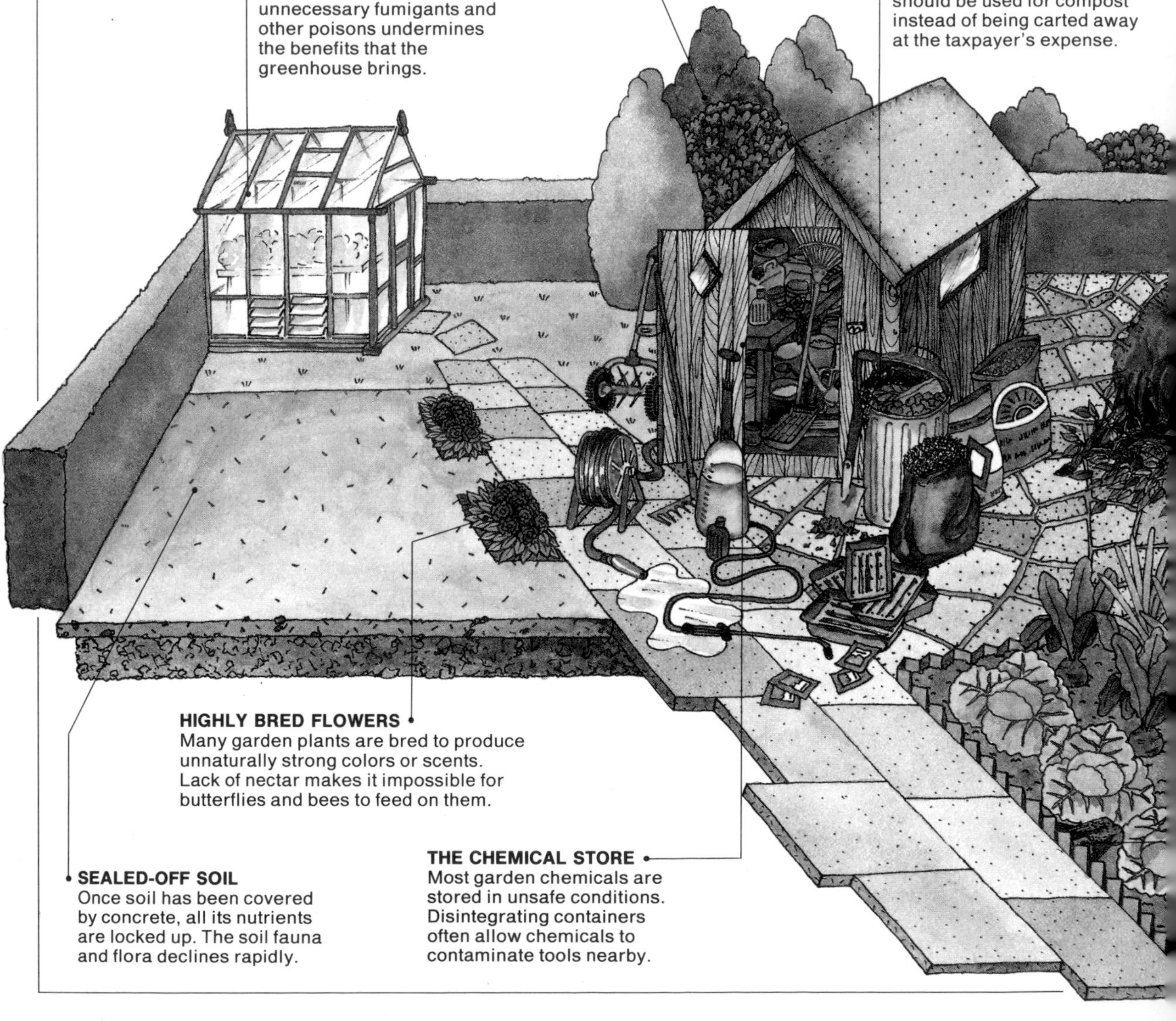

GREENHOUSE
Gardening under glass is a thoroughly productive activity. But filling a greenhouse with unnecessary fumigants and other poisons undermines the benefits that the greenhouse brings.

HIGHLY BRED FLOWERS
Many garden plants are bred to produce unnaturally strong colors or scents. Lack of nectar makes it impossible for butterflies and bees to feed on them.

SEALED-OFF SOIL
Once soil has been covered by concrete, all its nutrients are locked up. The soil fauna and flora declines rapidly.

THE CHEMICAL STORE
Most garden chemicals are stored in unsafe conditions. Disintegrating containers often allow chemicals to contaminate tools nearby.

POSITIVE ACTION

Gardening for food and wildlife

- **Encourage pollinators and predators**
  If you plant wild flowers these will attract pollinating bees, which will improve your fruit yields, and also hoverflies, which will feed on aphids.

- **Plant native trees and shrubs**
  Grow plants that are native to your area rather than ones that have been imported. They will be of greater value to birds and insects.

- **Keep your garden varied**
  Variety is the keynote when attracting wildlife and keeping down pests. Don't grow too many plants of the same type close together – try interplanting one crop with another.

- **Build a pond**
  Building a pond will attract frogs and toads into a garden. These animals are invaluable predators of slugs, snails, and ground-dwelling insects.

- **Avoid synthetic pesticides**
  If you do have pest problems, try using natural pesticides (see p. 135) or natural plant controls (see p. 138).

- **Recycle organic waste**
  Never throw away or burn organic matter – put it on the compost heap instead. If your soil is very poor, use organic fertilizers made from animal manure or plant matter (see p. 135) – never synthetic products.

REGIMENTED VEGETABLES
A soldierly vegetable plot filled with identical and unblemished plants is more often a sign of chemical contamination than a healthy crop.

PESTICIDE POISONING
Any animal killed by a pesticide concentrates that chemical in its body and then passes it on up the food chain. Birds that eat poisoned slugs and snails are often pesticide victims.

the onion-fly or their potatoes ruined by blight. But if we are not to turn our garden paradises into garden poison-patches, we need a completely new approach to the subject.

The orthodox approach sounds so simple, and it is – but it is too simple. Mankind has been able to bring to extinction many species of living things on this planet, but so far it has not succeeded in annihilating one single species of insect pest. DDT nearly annihilated the anopheles mosquito on the island of Sri Lanka, but when it was discovered that the DDT was doing the human population far more harm than the mosquitoes had done, its use had to be discontinued. Both mosquitoes – and malaria – came back far more virulently. Many insect pests are far worse problems to mankind on this planet now than they were before modern insecticides were invented. In spite of 40 years of constant chemical warfare all over the planet, more insect damage is being done than ever before.

STRATEGY INSTEAD OF CHEMICALS

The sensible approach to gardening, the organic approach, incorporates a recognition that we humans have to coexist with other forms of life on this planet. All living things are interdependent. We will never completely eliminate pests but we can control them when they get out of balance. But in order to control them, we must study them and we must know what we are doing to them.

Very often the modern orthodox gardener reacts with unnecessary violence before he has really been hurt. He cannot bear to see the least trace of damage to any of his crops. Very often, if he waited a little while, the predators which exist naturally for every insect pest would increase their numbers and do the job for him. After all, what does it matter if a few plants out of the many are slightly damaged by an insect pest every now and then?

But when a pest or a disease gets right out of control, and there is a risk of a serious crop loss, you must do something about it. There are a number of courses of action open to you that have nothing to do with chemicals.

For example, if you sow your broad bean seeds in the autumn instead of the spring, the beans will have made their full growth before the inevitable black aphid attack in the early summer. If you simply then pick the tops off (and either eat them or put them in the compost heap) you will encourage the beans to pod and discourage the aphids. No spraying whatsoever is required.

If you encourage birds, particularly tits, in your garden, you will eliminate many harmful insects in the egg stage during the winter. If you are careful about rotating your vegetable crops, so that you do not plant the same crop on the same land two years running, you will make things difficult for pests.

PEST CONTROL BY INTERPLANTING

Interplanting is a simple technique which makes full use of nature's own methods of pest control. It works because the natural characteristics of a number of plants will deter certain mammal and insect pests. Savory, for example, planted near beans will help ward off mice. Carrots interplanted with onions help to prevent both carrot- and onion-fly because each "jams" the scent of the other and this confuses the pests. For more examples of interplanting see the panel on page 138.

Certain plants, such as nasturtiums or convolvulus, encourage hoverflies which will then destroy many pests. In Britain alone there are hoverflies, ichneumon flies, braconid flies, and chalcid wasps – all of which destroy caterpillars and aphids or other pests. Ladybugs are voracious devourers of aphids. They lay their smooth yellow eggs on cabbage leaves, among other places, often right next to the yellow caterpillar eggs. When you destroy the latter, it is most important not to destroy the former. By blanket spraying with indiscriminate poisons you may wipe all these out and ultimately do only harm to your garden.

Handpicking of pests is not an activity to be despised. If you carefully examine the leaves of brassica (cabbage tribe) plants and simply squash the clumps of yellow caterpillar eggs you will destroy them just as surely as the most virulent poison would. Adult caterpillars too can be easily squashed. Cabbage white butterflies should be caught and destroyed regardless of whether the children think they look pretty or not. After all, it is better to destroy the butterflies than to risk poisoning the children.

THE GARDENER'S ALLIES

Many of the insects and animals that we spend so much time, money, and energy attempting to remove from our gardens are in fact beneficial. They are part of the process by which the balance of nature is maintained. This table shows some common animals that far from being a nuisance can help to eliminate pests.

INSECTIVORES

The animals in this group, which include shrews, moles, and hedgehogs, eat large numbers of invertebrate pests such as woodlice, millipedes, wireworms and slugs.

BIRDS

Many species of birds feed on grubs, caterpillars, slugs, and aphids, while the mistle thrush will account for large quantities of snails.

FROGS AND TOADS

Frogs and toads are important predators of slugs, woodlice, and other small insects. They are often found in damp grass, so care is needed when cutting it.

INSECTS

Both larvae and adult ladybugs, and the larvae of hoverflies and lacewings, are important predators of aphids.

GROUND-DWELLING INVERTEBRATES

Ground beetles and centipedes feed on eelworms, cutworms, leatherjackets, and other insect larvae. Spiders are devourers of insect pests.

EARTHWORMS

Earthworms help to aerate the soil and keep it well drained. They also improve soil fertility by carrying organic matter underground.

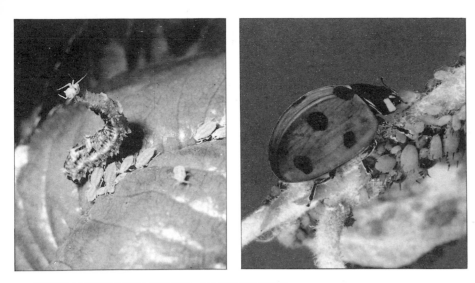

Nature's aphid controls
If you look closely at an aphid colony, you will probably see predators like ladybugs (*left*) and hoverfly larvae (*far left*) making short work of them. They are nature's aphid controls. Unfortunately, they are highly vulnerable to commercial aphid sprays. So it is worth remembering that soaking them with soft soap instead will harm the aphids but *not* their predators.

HOW PESTICIDES WORK AGAINST THE GARDENER

Before the era of modern pesticides, there were magnificent gardens. Any gardener over the age of sixty will remember superb crops of fruit and vegetables, and gardens – of both flowers and food – burgeoning with health and fertility. *None* of the organochlorines or organophosphates or carbamate pesticides existed in those days. Some gardeners made a limited use of sulfur: people learned to prevent blight on potatoes or grapes with mixtures of copper sulfate and lime. Nicotine, derris, or pyrethrum might have been used occasionally but all these things were nonpersistent and very often spared predators and only killed the target species of insect. Many gardeners hardly sprayed at all. One had "bad years for this" and "bad years for that" but somehow the balance of nature always reasserted itself and one always got a perfectly adequate crop in the end.

Now however, gardeners, like farmers (see pp. 44–6), have got themselves caught up on the pesticide treadmill. In his book *Fruit Pest and Disease Control,* the organic gardening expert Lawrence Hills describes how the pesticide treadmill was established in the case of the red spider mite on apple trees in Britain.

Ever since the Romans brought cultivated apples over to the British Isles, the red spider mite had lived harmlessly on apple trees eating mostly algae and lichens. The amount of harm it did to the trees was negligible. Then, in the early part of this century, the practice of tar washing was invented. This killed the algae and lichens but did not kill the red spider mites. So the mites were forced to turn to the apple foliage for their livelihood – and consequently became a pest. After World War II DDT was introduced, which destroyed all the mite's predators. The problem got worse. Now the red spider mite is a major pest, has developed resistance to all the organochlorine and organophosphate pesticides, and all attempts to control it seem to have failed. Fortunately though, one of its predators (the *Typhlodromus* mite) has begun to develop resistance to pesticides and so perhaps the situation will be saved after all – not *by* pesticides but *in spite of* them.

Gardeners, like farmers, have climbed up on to the pesticide treadmill because they have only considered each problem in isolation. In dealing with life, the simplistic answer is never the right one. Nor is treating symptoms ever the right answer – we must find the cause.

Overdressing crops with nitrates and other artificial fertilizers often leads to debility (plants, like children, can "outgrow their strength"). Growing crops in soil devoid of natural humus will lower resistance to disease. Plant breeders are laying up trouble for the future by breeding only for yield and appearance – and neglecting the plant's ability to resist disease. Potato breeders, for example, have used chemicals to protect their crops from blight for a hundred years now. The result is that modern potato varieties have lost all resistance. If they had had no blight protection during all that time, only blight–resistant varieties would have survived and spraying against blight would be quite unnecessary. Potatoes were one of the first crops to be protected chemically, but the same thing is happening nowadays with every known cultivated plant.

Now, dousing plants with chemical pesticides does other things besides encouraging pests to build up their natural resistance (so that sooner or later they will become even *more* of a problem than they were originally). Using chemicals *also* has appalling consequences on the condition of the soil: the sprays and powders that are aimed at your rose bush or cabbage leaves will eventually end up in the ground. They will either fall directly on to the soil and from there work their way into it, or they will be absorbed by the plant itself and sometimes reach the ground that way. And however they get there, the effect is the same: the essential balance of bacteria is destroyed and with it goes the rich source of minerals needed for successful growing. Once the soil in your garden is debilitated in this way, you can very easily get caught up in the trap of using other chemicals to try to liven it up again. But they don't; they just further aggravate the situation.

HOW PESTICIDES HARM OUR HEALTH

In recent years pesticides have been at the center of a raging controversy about the effect that these chemicals are having on our health. But little has been done to act on the findings of research into the subject – after all, governments make a lot of money out of the companies who produce the chemicals, so they are not likely to be quick to restrict their activities, are they? Nevertheless the evidence is clear: pesticides are not only toxic to plants and pests, they are toxic to humans too. They can cause cancer, birth defects, and genetic abnormalities. Organophosphates can reduce the liver's ability to detoxify the blood and carbamates can seriously harm the nervous system. There are eighty-nine pesticides currently on the British market alone that are known allergens and irritants. Yet the more chemicals we use, the more dependent we will get on chemicals – so we end up having to use still *more* of them. This may be good

Dressed to kill
Few people would consider wearing protective clothing for gardening, but people who deal with large commercial crops of flowers or plants certainly do. The chemicals used in pesticides and herbicides can have harmful effects on anyone. Although they are recognized as an occupational hazard, the risks to home-users are seldom adequately spelled out.

news for the chemical companies, but it is very bad news for our children and grandchildren.

STARTING AN ORGANIC GARDEN

Let us suppose that you have been gardening with chemicals, but wish to take up organic gardening instead. How do you go about it?

Well, the first and most obvious step is the total avoidance of synthetic chemicals – both biocides and artificial fertilizers. This does not rule out the use of natural pesticides, of which there are a fair number. In spite of your vigilance, or perhaps because of your lack of it, the enemy may sometimes get through, and then you may occasionally have to resort to using sprays.

None of the natural chemicals shown in the panel below are systemic (remaining within the tissues of plants) and none of them are very persistent. Most of them will act on the troublesome pest, without harming the pest's predators too much.

The most important part of any garden is the food-growing part, because what you put on *that* may end up on your plate. True organic gardeners do not use any herbicides on food crops: this prevents potentially dangerous chemicals becoming

Organic gardening To a chemicals-and-concrete gardener, an organic garden may look unruly and unproductive. In fact the lush and abundant growth is a sign of good health and burgeoning wildlife.

POSITIVE ACTION
Using natural fertilizers and pesticides

NATURAL FERTILIZERS
- **Animal manure**
  This is the most traditional fertilizer of all and needs to be well rotted before it is applied to prevent it initially robbing the soil of nitrogen.
- **Liquid plant manure**
  This can be made by packing the leaves of green manure plants such as comfrey into a water barrel. After a month a highly nutritious (if smelly) liquid is produced.
- **Bonemeal**
  This commercial organic fertilizer is rich in phosphate which is released slowly into the soil.
- **Hoof and horn**
  Another commercial organic product; it contains as much nitrogen as synthetic fertilizers.
- **Wood ash**
  This is the gardener's best source of potash – quite as effective as mined potash fertilizers.

NATURAL PESTICIDES
- **Soft soap**
  This semi-liquid soap kills aphids by dissolving the wax layer on their skins so they die of desiccation.
- **Nicotine**
  This *highly poisonous* but natural insecticide kills aphids, scale insects, and caterpillars but not ladybugs or hoverflies. It decomposes naturally after use.
- **Quassia**
  This is a natural insecticide found in the bark of the tropical quassia tree. The bark is sold in dried form: soaking it produces quassia spray.
- **Potassium soap**
  Another harmless aphid control. The soap is dissolved in water and then used as a spray.
- **Derris-pyrethrum**
  A plant-produced insecticide that is toxic to nearly all insects – it must be used with care.

incorporated in food – and those who eat it.

Once you have stopped using synthetic chemicals, the next step is to improve the fertility of the soil.

RESUSCITATING THE SOIL

Soil that has become completely hooked on chemicals may take some time to recover, for three main reasons. One is that the content of organic matter in it will be low. The second is that the numbers of nitrogen-fixing bacteria will also be low. The third is that the number of beneficial living organisms such as predators will be low too. So in an organic garden these things must be built up again. All three problems can be solved by applying compost to the soil. This will increase the organic matter in the soil, increase the level of nitrogen-fixing bacteria and, as a result of healthy soil unpolluted by chemical poisons, the predators will gradually return too.

MAKING A COMPOST HEAP

So the organic gardener must build a compost heap. This should be a neatly built stack of organic matter. Ideally, the compost heap will be contained within walls made of brick or concrete blocks, or plank fencing with ventilation holes. Sticks or other coarse material are placed at the bottom, to let the air in. Nearly all vegetable waste from the garden – or from other sources – can be used and is then laid in orderly layers inside the construction. Lawn mowings can be mixed with other, coarser material to ensure that it rots properly. Kitchen waste – even fish-heads or bits of meat – can be incorporated in the compost heap (ignore the advice of those who say they can't). But it needs to be placed right at the heart of the heap and buried deeply, so that the cats and dogs can't get at it, and it must be allowed to decay thoroughly before the compost is used.

By building a second compost heap alongside the first, the organic gardener will find that the first has rotted down and is ready to use by the time the second is built. Spreading this compost on your garden is entirely beneficial and will do only good to everything.

The burgeoning fertility of a true organic garden can provide nearly all of the material needed for the heap. Organic material for the heap is not difficult to come by outside the garden: leaf-mold from the neighboring park or woods, seaweed from a visit to the coast, leaf-sweepings from the streets in the autumn which you can generally get free, some of

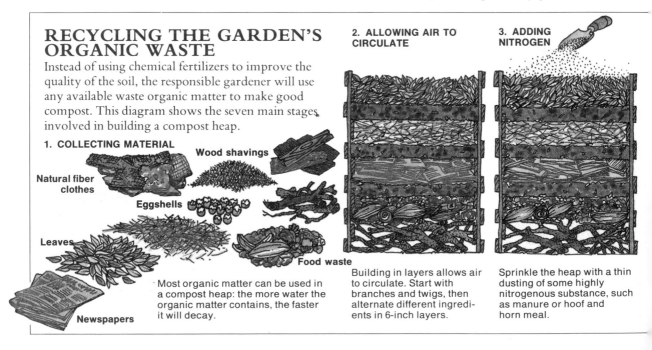

RECYCLING THE GARDEN'S ORGANIC WASTE

Instead of using chemical fertilizers to improve the quality of the soil, the responsible gardener will use any available waste organic matter to make good compost. This diagram shows the seven main stages involved in building a compost heap.

1. COLLECTING MATERIAL

Wood shavings

Natural fiber clothes

Eggshells

Leaves

Food waste

Newspapers

Most organic matter can be used in a compost heap: the more water the organic matter contains, the faster it will decay.

2. ALLOWING AIR TO CIRCULATE

Building in layers allows air to circulate. Start with branches and twigs, then alternate different ingredients in 6-inch layers.

3. ADDING NITROGEN

Sprinkle the heap with a thin dusting of some highly nitrogenous substance, such as manure or hoof and horn meal.

the vast amount of rotting vegetables that all green-grocers throw away every day, horse manure from the local riding school. Every true organic gardener develops an eye for these things and would probably find something for the compost heap even in the middle of the Sahara Desert.

BUILDING UP BACTERIA

The nitrogen that is so precious to the soil is incorporated into the compost heap as the heap is rotting. It is absorbed from the air by bacteria attached to the organic material. But before this process of fixing the nitrogen can begin, the organic gardener has to "activate" the compost heap, in order to get the rotting going: the compost heap needs to be very hot if it is to work properly. This heat will kill all weed seeds, harmful micro-organisms, and other undesirables. Young green growth has enough nitrogen in it to achieve this heat but any dry matter such as straw does not.

You must therefore add some nitrogen. A sprinkling of any highly nitrogenous substance will do the trick – a good dusting of it over 6 or 8 inches of plant material. Organic purists insist on organic nitrogen such as fish meal, blood meal, animal or human excreta, or urine. But there is no need to have any qualms at all about using so-called "inorganic" nitrogen for this purpose if "organic" nitrogen is hard to obtain. Any highly nitrogenous fertilizer which you can buy expensively from the garden center, or much more cheaply from the agricultural merchant, will do.

GREEN MANURING

Another tool for upgrading the organic content of your soil is "green manuring." Plants of the pea family are an excellent crop for this purpose. Such crops are dug in, or plowed or rotated in. Alternatively they are pulled up and put on the compost heap at the flowering stage before they have begun to seed and when they are still young and green and full of nitrogen.

Mustard works well as a quick green manure crop. It is not nitrogen-fixing but it provides a lot of bulk very quickly. Hungarian rye is invaluable for a winter cover. None of the pea family, with the possible exception of winter tares, grow much in the winter, but any of the following can be sown in the early autumn, and may well survive the winter: vetch, rough peak sour clover, crimson clover, bur clover, Austrian winter pea. These crops can also be dug in during the spring.

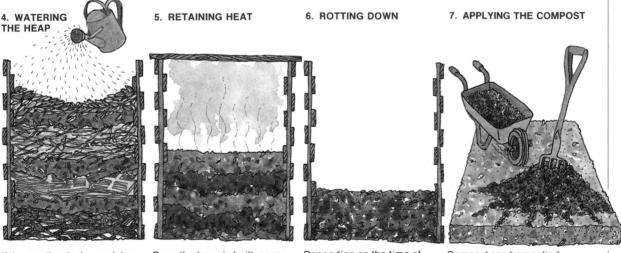

4. WATERING THE HEAP

If the weather is dry, moisten the heap by sprinkling lightly with water. This can be done with each layer if the material is very dry.

5. RETAINING HEAT

Once the heap is built, cover it with a layer of earth and a lid of some sort to keep the heat in and speed up the rotting process.

6. ROTTING DOWN

Depending on the time of year and the weather, a 5-foot high heap will take between 3 and 6 months to rot down.

7. APPLYING THE COMPOST

Compost can be applied directly to the soil or dug in, depending on the time of year.

POSITIVE ACTION

Using plants to control pests and disease

Many plants have chemical defenses against pests and disease. By growing them in strategic places, or by mixing them with other plants, you can help to reduce the impact of weeds and animal pests in your garden without chemicals.

- **Herbs**
  The fragrance of many herbs such as sage, rosemary, thyme, and peppermint will deter cabbage butterflies and slugs. Lavender provides some defense against ants and helps protect roses against aphids.

- **Wormwood**
  The odor of wormwood is repellent to many insects. Known also as absinthe, it is traditionally planted near currant bushes to protect them.

- **French marigold**
  This attractive flower can be used to repel a variety of the pests of vegetables. Try interplanting them with cabbages or tomatoes.

- **Garlic**
  The germicidal and fungicidal properties of garlic work as well in the garden as they do in the human gut. Planting garlic between strawberries, next to roses and beneath fruit trees can be very effective against fungus diseases.

- **Onions**
  Onions, like garlic, have a fungicidal effect and can usefully be grown between strawberry plants to discourage diseases. Planted beside carrots, onions will also deter carrot-fly.

Another essential technique of organic gardening is mixed cropping (growing several species of plant at a time) and rotation. The principle behind this is not to grow the same crop on the same ground for two years running, but to leave several years between crops of the same sort. This technique avoids the problems of debilitated soil which in turn encourages the use of chemicals in an attempt to counteract them.

GARDENING FOR YOUR GRANDCHILDREN

To describe all the techniques and merits of organic gardening would require a whole book. There are a number of associations (see p. 185) which will encourage and advise gardeners who wish to turn to organics. Once you have mastered the principles involved, and are gardening with the due respect for Nature and ourselves, you may be left exhausted from honest hard work and fresh air, or even with a touch of backache, or a few blisters, but your activities will not have involved the use of chemicals that endanger the future of our planet – whether by upsetting the natural balance of pests and plants, or by causing cancer and deformities. Organic gardening is in accord with the principles of good housekeeping. The principle of moderation ensures that pesticides – even natural ones – are not used unnecessarily. Since other creatures have as much right to live as we have, the principle of nonviolence should guide us towards ways of protecting our food supply as moderately and benignly as possible. Every gardener should beware of killing forms of life that are not hurting them. The principle of responsibility should make us ask ourselves whether we have a right to put down, for example, poisonous slug pellets if these are likely to be taken up by the harmless – indeed beneficial – birds? Maybe they won't harm the birds – but it is our business to know. And it is not good enough just to believe "what it says on the label."

Gardening without chemicals is a way of benefiting from nature without harming it. Once rid of these substances, your garden will become, if not paradise itself, at least free of the poisons that permeate so much of the rest of the Earth. It is these chemicals that are impairing the quality of our lives on this planet now, and that give it a far from pleasing prospect for the future. But if we care about the quality of our lives, then we must take action right away. Throwing out the pesticides and herbicides, the fungicides and fertilizers, won't instantly solve the problems of worldwide pollution, nor will it solve the problems caused by stupid politicians and greedy businessmen. But it will create a small chink in the armour of the chemical market – and a little bit of poison-free planet for you and your children to enjoy.

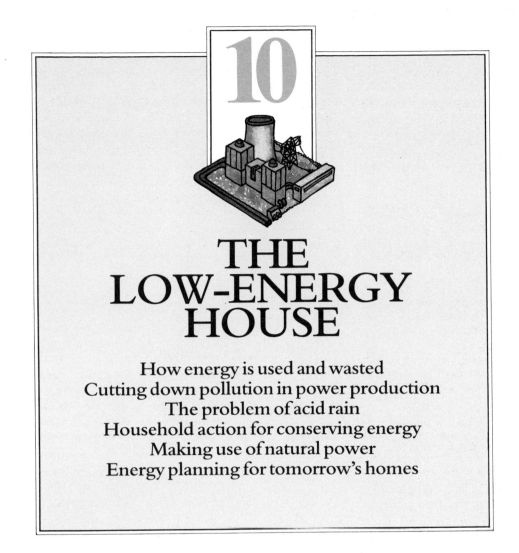

10

THE LOW-ENERGY HOUSE

How energy is used and wasted
Cutting down pollution in power production
The problem of acid rain
Household action for conserving energy
Making use of natural power
Energy planning for tomorrow's homes

When most of the television viewers of an entire country get up from their armchairs during a commercial break in *Dallas,* to relieve their bladders and turn on the kettle, a huge chain reaction is triggered in the energy supply system. Millions of horsepower of electricity are conjured up at the drop of a hat.

Well, not exactly the drop of a hat. Because tucked away in the control rooms of most electricity authorities there is someone whose job it is to read the paper. Not any part of the paper, but the day's television schedules. So when the commercial break begins, the electricity people are ready for it.

As millions of switches are flicked on, a huge amount of current is drained from the supply. To provide this, more power is produced from generators, so more steam has to be on hand to speed them up. To produce that steam, extra heat is supplied from furnaces, so more coal, oil, or gas must be made available to provide that heat. As the extra fuel is burned, so more smoke, and with it, more poisonous gases, pour out from the tall chimney stacks. In nuclear power stations, thousands of the control rods which damp down the reactors are inched out so the nuclear reaction "burns" more brightly. As this happens, the radioactive fuel in the reactors takes another step towards the time when it will have to be replaced.

But how many people realize what effect this synchronized switching on of kettles has? How many even think about where the electricity comes from when the switch is flicked, and how it is produced? The answer is very few: nearly everyone takes it completely for granted.

THE ILLUSION OF CLEAN POWER

The electricity industry is fond of calling its product "clean power". As they correctly point out, modern all-electric houses are not enveloped by palls of smoke, electricity does not choke whole neighbor-

hoods with sulfurous fumes, and the streets of cities like London are no longer blackened by soot or shrouded in smoke.

But although your house may not have a smoking chimney itself, somewhere over the horizon a power station is probably pouring out smoke on your behalf. And it will be doing so day and night, all the year round, in order to produce electricity for factories, offices and perhaps a million other fuel-hungry households as well as your own.

THE ENERGY CHAIN
Part 1: How energy is produced

In all developed countries, the production of energy is highly centralized. Power stations may be far removed from the places where their energy is consumed, but the pollution they cause travels far and wide and affects air, sea, and land.

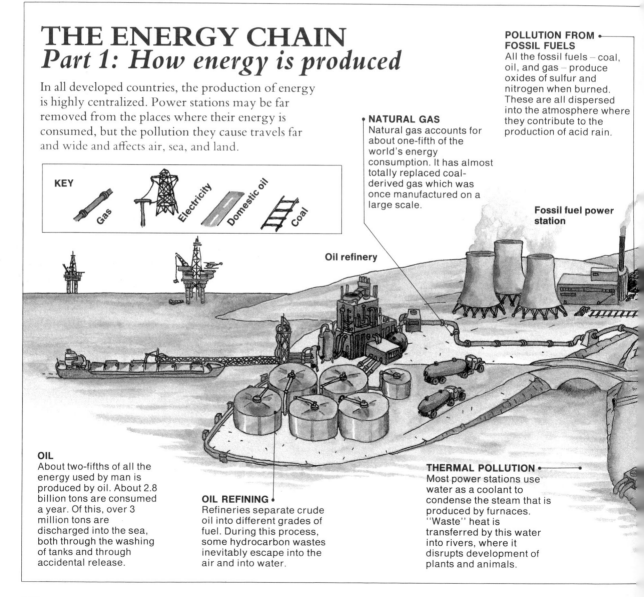

KEY
Gas
Electricity
Domestic oil
Coal

POLLUTION FROM FOSSIL FUELS
All the fossil fuels – coal, oil, and gas – produce oxides of sulfur and nitrogen when burned. These are all dispersed into the atmosphere where they contribute to the production of acid rain.

NATURAL GAS
Natural gas accounts for about one-fifth of the world's energy consumption. It has almost totally replaced coal-derived gas which was once manufactured on a large scale.

Fossil fuel power station

Oil refinery

OIL
About two-fifths of all the energy used by man is produced by oil. About 2.8 billion tons are consumed a year. Of this, over 3 million tons are discharged into the sea, both through the washing of tanks and through accidental release.

OIL REFINING
Refineries separate crude oil into different grades of fuel. During this process, some hydrocarbon wastes inevitably escape into the air and into water.

THERMAL POLLUTION
Most power stations use water as a coolant to condense the steam that is produced by furnaces. "Waste" heat is transferred by this water into rivers, where it disrupts development of plants and animals.

Power station chimneys are up to 1,000 feet high – about as high as the Eiffel Tower. They are built like this because the engineers who designed them imagined that they would disperse the fumes over a large area, far above towns and cities, and so render them harmless. Disperse them they did, but harmlessly they did not. It is now all too obvious that the plan has backfired: the tall chimneys of power stations are now spreading environmental havoc on an unprecedented scale.

ACID RAIN AND HOUSEHOLD ENERGY

Anyone who has been in a room where a coal fire was burning will have breathed in sulfur dioxide. It is sharp and acrid – not dangerous in small quantities perhaps, but thoroughly eyewatering when the wind blows the smoke back down the chimney. This choking gas pours out of all coal-, oil-, and gas-fired power stations. Worldwide around 100 million tons of the stuff are now released annually, making sulfur dioxide the

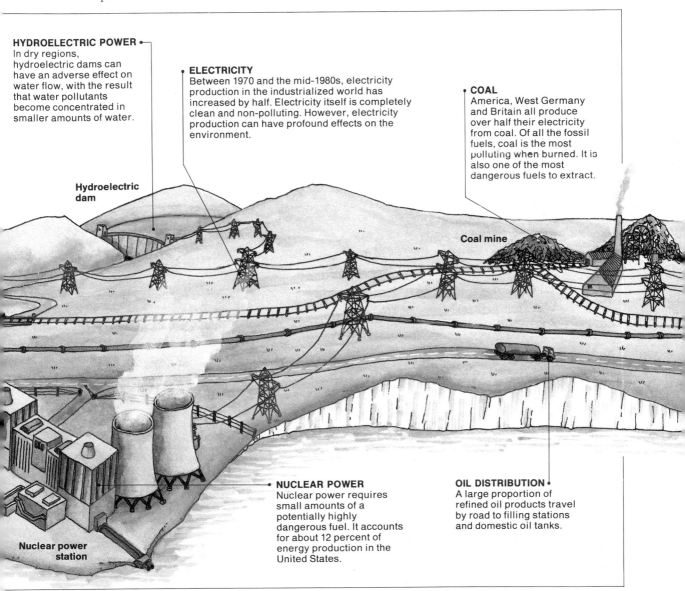

HYDROELECTRIC POWER
In dry regions, hydroelectric dams can have an adverse effect on water flow, with the result that water pollutants become concentrated in smaller amounts of water.

Hydroelectric dam

ELECTRICITY
Between 1970 and the mid-1980s, electricity production in the industrialized world has increased by half. Electricity itself is completely clean and non-polluting. However, electricity production can have profound effects on the environment.

COAL
America, West Germany and Britain all produce over half their electricity from coal. Of all the fossil fuels, coal is the most polluting when burned. It is also one of the most dangerous fuels to extract.

Coal mine

Nuclear power station

NUCLEAR POWER
Nuclear power requires small amounts of a potentially highly dangerous fuel. It accounts for about 12 percent of energy production in the United States.

OIL DISTRIBUTION
A large proportion of refined oil products travel by road to filling stations and domestic oil tanks.

greatest of all man-made environmental hazards apart from radioactive waste.

The tall chimneys disperse these sulfur-laden fumes over huge areas, often far beyond national boundaries. They get carried across land and sea by the prevailing winds. Close to their source some of the particles fall to the ground as "dry deposition," while over longer distances – often many hundreds of miles – the sulfur dioxide dissolves in water vapor to produce sulfuric acid, and falls to the ground as the now-notorious acid rain.

There can be few global calamities that have appeared with so little warning but that have had such disastrous and wide-ranging effects. The Scandinavians were the first to draw the attention of the European public to acid pollution at a United Nations conference in 1972. When Swedish scientists presented an initial report it aroused disbelief: it suggested that sulfurous emissions from power stations and factories in countries like West

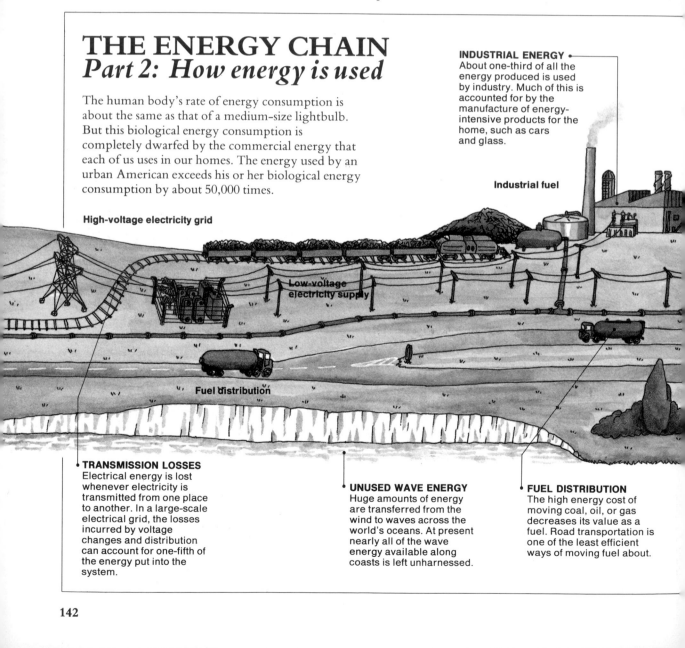

THE ENERGY CHAIN
Part 2: How energy is used

The human body's rate of energy consumption is about the same as that of a medium-size lightbulb. But this biological energy consumption is completely dwarfed by the commercial energy that each of us uses in our homes. The energy used by an urban American exceeds his or her biological energy consumption by about 50,000 times.

INDUSTRIAL ENERGY
About one-third of all the energy produced is used by industry. Much of this is accounted for by the manufacture of energy-intensive products for the home, such as cars and glass.

Industrial fuel

High-voltage electricity grid

Low-voltage electricity supply

Fuel distribution

• TRANSMISSION LOSSES
Electrical energy is lost whenever electricity is transmitted from one place to another. In a large-scale electrical grid, the losses incurred by voltage changes and distribution can account for one-fifth of the energy put into the system.

• UNUSED WAVE ENERGY
Huge amounts of energy are transferred from the wind to waves across the world's oceans. At present nearly all of the wave energy available along coasts is left unharnessed.

• FUEL DISTRIBUTION
The high energy cost of moving coal, oil, or gas decreases its value as a fuel. Road transportation is one of the least efficient ways of moving fuel about.

Germany and Britain were causing colossal environmental damage in far-away Scandinavia, wiping out fish stocks in many lakes and in some areas making water dangerous for human consumption. In the accused countries the news was greeted with scepticism or even hostility.

Today the disbelief has gone. Over twenty thousand lakes in Sweden alone have been killed by acidification. Many thousands more in Norway, Canada, and Scotland have suffered the same fate.

In much of the northern hemisphere, forests have been the next victims. If you get in a car and drive around Europe you can see the appalling effects that this is having. Everywhere there are dying trees. Firs in the Black Forest, in Bavaria, or the Harz mountains in Germany are now endangered species. Those that have survived have strange flattened tops, and their branches are miserably bare most of the way down the trees' trunks. Spruce trees throughout Europe have suffered from

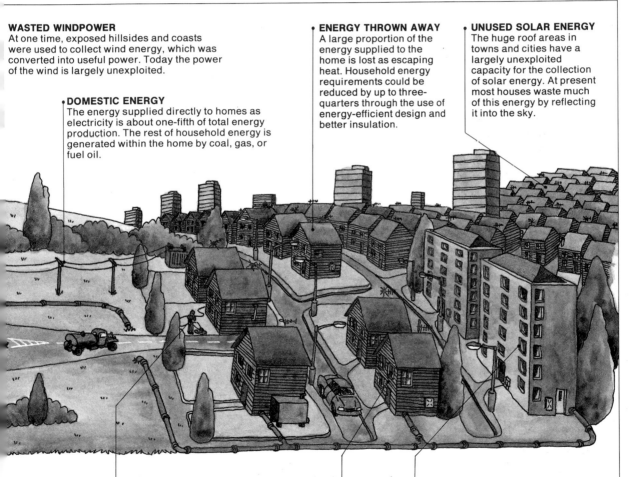

WASTED WINDPOWER
At one time, exposed hillsides and coasts were used to collect wind energy, which was converted into useful power. Today the power of the wind is largely unexploited.

DOMESTIC ENERGY
The energy supplied directly to homes as electricity is about one-fifth of total energy production. The rest of household energy is generated within the home by coal, gas, or fuel oil.

ENERGY THROWN AWAY
A large proportion of the energy supplied to the home is lost as escaping heat. Household energy requirements could be reduced by up to three-quarters through the use of energy-efficient design and better insulation.

UNUSED SOLAR ENERGY
The huge roof areas in towns and cities have a largely unexploited capacity for the collection of solar energy. At present most houses waste much of this energy by reflecting it into the sky.

HOUSEHOLD POWER
Power is used in the home by electrical machinery such as washing machines, dishwashers, vacuum cleaners, and lawn mowers. Of all appliances, freezers and stoves consume the most energy per year.

ENERGY FOR HEATING
Heating homes and water uses large amounts of energy. To keep a 2-kilowatt electric heater running for just three hours a day consumes the equivalent of half a ton of oil a year.

ENERGY FOR LIGHTING
Lighting accounts for a modest part of household energy consumption. A single 100-watt lightbulb, if switched on for eight hours a day for a year, will consume the energy produced by about 150 pounds of oil.

ACID RAIN AND THE DEATH OF FORESTS

The precise way in which atmospheric pollution kills trees is tangled in complexity. But most researchers agree that the great increase in man-made aerial pollutants is responsible for this catastrophe, even if by very indirect routes. This diagram shows some of the ways by which air is polluted and how the atmospheric changes it causes may kill trees.

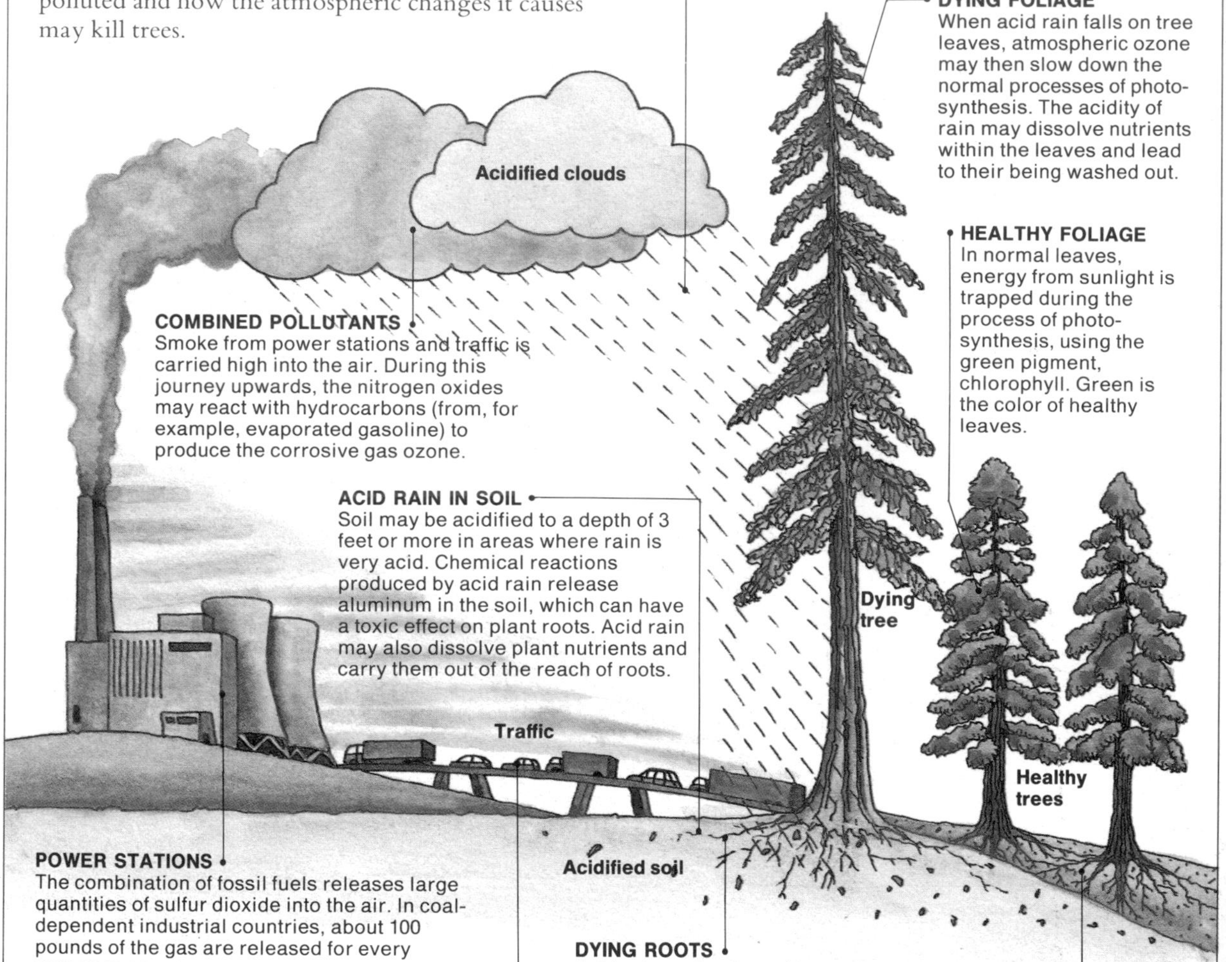

ACID RAIN
When they meet moist air, nitrogen oxides and sulfur dioxide dissolve to produce sulfuric and nitric acids. The clouds become strongly acidified, and these produce acid rain.

DYING FOLIAGE
When acid rain falls on tree leaves, atmospheric ozone may then slow down the normal processes of photo-synthesis. The acidity of rain may dissolve nutrients within the leaves and lead to their being washed out.

HEALTHY FOLIAGE
In normal leaves, energy from sunlight is trapped during the process of photo-synthesis, using the green pigment, chlorophyll. Green is the color of healthy leaves.

Acidified clouds

COMBINED POLLUTANTS
Smoke from power stations and traffic is carried high into the air. During this journey upwards, the nitrogen oxides may react with hydrocarbons (from, for example, evaporated gasoline) to produce the corrosive gas ozone.

ACID RAIN IN SOIL
Soil may be acidified to a depth of 3 feet or more in areas where rain is very acid. Chemical reactions produced by acid rain release aluminum in the soil, which can have a toxic effect on plant roots. Acid rain may also dissolve plant nutrients and carry them out of the reach of roots.

Dying tree

Traffic

Healthy trees

POWER STATIONS
The combination of fossil fuels releases large quantities of sulfur dioxide into the air. In coal-dependent industrial countries, about 100 pounds of the gas are released for every person every year.

Acidified soil

CARS AND TRUCKS
Traffic is a prime source of nitrogen oxides, of which there are a number of different types. Nitrogen oxides may persist in the atmosphere for up to two months.

DYING ROOTS
Acid rain on the ground, together with the acid rain that trickles down tree trunks, kills the fine hairs on tree roots. The deepest roots die back, leaving the tree vulnerable to strong winds.

HEALTHY ROOTS
In healthy trees, the root system reaches many feet below the surface. Water and minerals are absorbed by microscopic hairs which cover the roots' surfaces.

The price of acid rain

Acid rain damages both the living and the inanimate. This statue in Pennsylvania (*above*) is deeply eroded by acid rain; throughout Europe, trees (*right*) are being killed by it. In Sweden, lime is added to lakes (*below*) in a desperate bid to prevent all aquatic life perishing.

massive needle loss, and pitifully limp twigs hang off the branches. All over Britain, from the highlands of Scotland to the rocky Cornish coast, from the Lake District to East Anglia, there are beech trees whose branches have been mutilated and which now no longer resemble beech branches at all. Every spring, yew trees lose more and more of their needles. Compare old photographs of yew trees in churchyards with new ones, and you will be dismayed by how desperately bare they look now.

By 1986 half of West Germany's forests were suffering from *Waldsterben*, the forest-wasting disease first diagnosed in the early eighties. Switzerland, Austria, Holland, Sweden, Czechoslovakia, Poland, and Canada are all suffering from forest dieback, and warning signs have been found in many neighboring countries. In Scotland, one particular storm has produced rain more acid than malt vinegar.

In addition to sulfur dioxide, over three thousand gases are being released into the atmosphere by factories, power stations and also by cars. Many are present in tiny quantities but it is known that they can dramatically affect atmospheric chemistry. The most recent research on tree death suggests that there is no one cause – trees are probably dying from a hideous pot-pourri of pollution made up of dozens or hundreds of chemicals.

But whatever the chemicals involved, there is no doubt whatever that acid rain, acid lakes, and dying trees are all caused by two things: road transportation, which we shall deal with in the next chapter, and our use of energy – and electricity in particular – at home, in our factories and our public buildings. There is no such thing as clean power: the fumes blown out of the chimneys of our power stations are the bad breath of modern man.

CLEANING UP POWER STATIONS

It was an Englishman, Mr. F. Goodfellow, who first developed and demonstrated an apparatus for removing sulfur dioxide from smoke. He did this not a few years ago, nor even a few decades ago, but in 1880. He was concerned about the effect of acid smoke in industrial Manchester, and so he decided to do something about it. But his desulfurization apparatus never caught on. Instead it disappeared into obscurity, and Victorian factory owners who lacked Mr. Goodfellow's farsightedness continued to let more or less what they liked issue forth from their chimneys. Over a century later, we are having to come to grips with the legacy of this neglect.

Fortunately there are ways that this pollution can be stopped, but they are adopted only when consumers make sufficient fuss, for it is fuss that produces results.

The sulfur that disappears up chimneys is potentially very useful. Huge amounts of sulfur are dug up from the ground in one way or another every year by industry, and so it is all the more nonsensical to waste the same element by scattering it into the atmosphere where it does nothing but damage.

At one power station in West Germany, the sulfur that would otherwise have poured out of the chimneys has been turned to good use. In the early seventies, the local electricity supply company at Wilhelmshaven on the North Sea coast decided that more power was needed for the growing number of factories. But by then, the menace of acid rain was making itself known, and the proposals were met by angry demonstrations. The local inhabitants, quite understandably, did not want to be showered with pollutants.

So the people of Wilhelmshaven were rewarded with a different kind of power station. It was the first full-scale coal- and oil-burning power station in Europe to be fitted with desulfurization equipment (at a time when the company was under no legal obligation to do so). When it started to produce electricity in 1977, it also began to test methods of cleaning up other waste gases which have since been fitted to other power stations.

Today the Wilhelmshaven power station not only generates electricity but also produces gypsum. This is created by spraying the smoke from the furnaces with lime. Now gypsum is an ingredient in the manufacture of concrete, and can be sold to cement factories. So the sulfur that would have ended up in the air is used up harmlessly in a product that has little impact on the atmosphere.

In the process used at Wilhelmshaven over nine-tenths of the sulfur is removed from the flue gases. This increases the cost of generating electricity by a trifling 10 percent. There is no doubt that consumers everywhere could absorb this added cost by using energy-efficient household appliances and by a more sparing use of all electrical equipment. It would be a small price to pay for preventing such major damage to our environment.

CLEANER WAYS TO BURN COAL

Of all the fuels we use, coal is probably the most messy to burn. But it needn't be – in fact, a lot can be done to prevent coal pollution *before* the smoke goes up the chimney.

First, much of the coal's sulfur – the substance responsible for the greater part of its polluting effect – can simply be washed out before the coal is burned. Of course, this costs money. A lot of coal is washed at present, but if more money were spent on this, the improvement would be immediate.

Second, the way coal is burned can be improved by reducing it to a powder, and then blowing it into the furnaces. In this "fluidized bed" system, which is being tested in a number of countries, each ton of coal produces more heat. Furthermore, because the furnaces do not reach extremely high temperatures, fewer noxious gases are produced. The result is more electricity but less pollution, in other words, a system that will make coal more acceptable until we can find something better to replace it.

THE NUCLEAR "ALTERNATIVE"

Not that many years ago, a leading figure in America's electricity industry confidently announced that nuclear power would make electricity meters a thing of the past. Electricity would be so plentiful that, like water, there would be unlimited supplies for everyone.

Well, he was wrong. Very wrong in fact. Nuclear power has not turned out to be cheap. But until early 1986 it might have been said to be *clean*. After all, nuclear power stations don't produce huge quantities of smoke, and they don't release sulfur dioxide, nitrogen oxides, or carbon dioxide into the atmosphere. Of course, the disposal of nuclear waste is no small matter, but it's a problem which the industry has become adept at explaining away or sweeping under the carpet.

Not even accidents like the Windscale fire in 1957 or the Three Mile Island near-disaster in 1979 dislodged the industry from its lofty perch. No one was killed and most of the damage was contained. Until April 1986 the nuclear industry could argue

that there was no definite proof that a single person had actually been killed by nuclear power while coal mining worldwide claimed a large number of victims every year.

The Chernobyl disaster changed all that: the worst fears of the anti-nuclear movement were suddenly confirmed. An invisible monster was on the loose, and within days the weather report on television in every European living room also be-

THE HIDDEN COSTS OF PRODUCING ENERGY

Producing energy always has a hidden cost: all of the "orthodox" methods of energy production result in hazards to the environment, to wildlife, and to human health. With the exception of hydroelectric power, which is only practicable in certain countries, all current large-scale methods have drawbacks – yet it is often the most dangerous of them which are used to produce the most power.

Method	Environmental hazards	Wildlife hazards	Health hazards
Nuclear power	Danger of release of radioactivity into air, water and soil. Unknown future risks from waste storage and old power stations.	Destructive effect of accidental radiation releases. Long-term hazard posed by the disposal of nuclear waste.	Unpredictable dangers from major accidents. Risk of cancer to people exposed to low-level radiation from nuclear waste.
Oil-fired power	Pollution from accidental oil spillages. Atmospheric pollution and acidification by waste gases from furnaces.	Mass destruction of marine life, from plankton to fish and birds, caused by oil spillages.	Risk of explosions from stored oil; possible risk from atmospheric pollution.
Coal-fired power	Dereliction of land by mining, especially with surface mines. Severe atmospheric pollution and acidification from impurities.	Poisoning of plants by mining waste; some risk to water life through waterborne pollution.	Indirect health hazard through atmospheric pollution, especially from sulfur dioxide. Considerable health hazards during mining.
Gas-fired power	Minor degree of air pollution as a result of burning.	Some destruction of habitats by pipelines, otherwise marginal effect.	Risk of explosions, otherwise little direct effect.
Hydroelectric power	No pollution hazards. Loss of land through flooding; minor risk of landslides.	Destruction of habitats through flooding. Disturbance of river life through altered water flow.	Safe, apart from danger of dam bursts.

came a radiation report. The nuclear cloud over Europe was all the more terrible for the uncertainty it brought with it. What effect would it have? Was it safe to go out? Should children be allowed to drink milk, or play in the street among possibly radioactive puddles?

Most of these questions will never be answered. Self-appointed "experts" have announced their findings, but how can anyone be expert in events that have never before been experienced? If a power station disaster hundreds, or even thousands, of miles away can cause dangers to health, contaminate soil, and render livestock which has grazed on that soil inedible, what might be the consequences of a serious accident in a *local* reactor?

As the environmental consequences of the Chernobyl disaster unfold, it may well turn out that the problems of acid rain are as nothing compared with the problems of radioactive rain. So can we justify the comfort of nuclear electricity if it brings with it such extraordinary dangers?

THE EVERYDAY HAZARDS OF NUCLEAR POWER

It is becoming apparent that even under normal operating conditions nuclear power can present major hazards to us all. Even if a power station runs safely, it produces dangerous waste which is reprocessed before being disposed of in some way. The factories that reprocess radioactive materials from nuclear reactors, such as Sellafield in Britain, are known to pose a potentially enormous danger to the environment and people's health. Since it was opened in the late fifties, this one nuclear complex has released at least 550 pounds of plutonium, the most lethal of all man-made poisons, into the Irish Sea — enough to give the entire population of Europe lung cancer if it was inhaled instead of submerged. Local leukaemia statistics point an accusing finger at the plant. But a similar processing site has come on stream in Cap de la Hague in France, and another is under construction at Wackersdorf in West Germany.

No one can claim with complete certainty that any of these plants (an unadvertised part of whose purpose is to produce weapons-grade plutonium) will not at some time in the future become major

Misplaced trust
Sizewell reactor (*above*), which has been linked with leukaemia, lurks behind local anglers.

Before the explosion
Chernobyl's reactor hall (*below*) looked as safe as any other two months before it blew up.

new sources of radioactive contamination.

We should not be fooled into thinking that nuclear power is a sensible alternative. Two countries, Austria and Sweden, have held referenda on nuclear power, and the people of both have decided that it wasn't for them. Austria has now mothballed its only completed nuclear power station which has never been in operation, while Sweden will phase out nuclear power when its twelve stations are eventually decommissioned because of "old age." In the United States no nuclear power stations have been commissioned since the Three Mile Island accident. In most countries operating nuclear power plants, public opinion is strongly in favor of getting rid of them.

We shall probably never know the full price of the Chernobyl disaster, because the Russians will not make it known. Indeed, they may never be able to work it out for themselves, because its effects will be too wide-ranging. But neither do we know the day-to-day dangers that our own nuclear power stations present. The risks of nuclear electricity are simply far too great for us to place any faith in it.

HOUSES THAT WASTE ENERGY

It is all very well to blame the power companies and the big industrialists for the pollution caused by producing energy. But they may reply, with some justification, that they are only supplying what we ask for. After all, from 30 to 45 percent of total energy production in Europe and North America ends up in the home. We have to make our homes more energy-efficient.

Unfortunately, the architects of the earlier postwar decades didn't care for the notion of energy conservation. They had lots of cheap energy and cheap glass, so up went houses with thin walls, vast windows, and lofts without any insulation. They designed buildings, in fact, that were exactly the opposite of traditional houses, with their thick walls and sensibly sized windows.

We are still paying the price for their handiwork. Half of the energy that goes into older houses is effectively thrown away. In a typical poorly insulated house, a third of this squandered heat goes through the roof, a quarter through the walls, up to a fifth through the floor, while a tenth is lost

YOUR YEARLY ENERGY USE

How much energy does a single person consume in a year? The answer depends very much on where you live. All the energy we use – whether from oil, coal, gas, hydroelectric, or nuclear sources – can be expressed as an "oil equivalent." This is the amount of oil that would be needed to supply the same quantity of energy. This diagram shows the oil equivalent, in standard barrels, for the energy each person consumes at home in a year in five different parts of the developed world.

Country	Barrels per person	
North America	22	
Australasia	16.5	
Scandinavia	13.5	
Europe	9.75	
Japan	8.5	

TOTAL ENERGY USE
Domestic energy consumption, as shown here, is only one-third of the total energy consumption per person. The other two-thirds are consumed in transportation, agriculture, and industry.

through drafts. Clearly there is no point going all out to reduce power station pollution if so much of the energy they produce is wasted.

Luckily the oil shocks of the seventies gave the architects a jolt. Electricity bills and heating costs shot up and so did the interest in insulation and the energy-conscious design of new houses. The environmental impact of our reckless use of energy has, quite rightly, become a source of concern.

ENERGY-SAVING AT HOME

The amount of energy spent on heating and cooling adds up to three-quarters of domestic energy use. The waste is colossal, but then so are the savings that can be made by proper insulation of roofs, walls, and floors, by the installation of double or even triple glazing and by draftproofing.

True energy-efficient houses are still a rarity, but prototypes do exist. The floors, walls, and ceilings are well insulated, while most have a whole-house ventilation and heat recycling system. "Active" solar hot water heaters are installed on the roof with "passive" solar air heaters on the south-facing side of the house. Even on cold, overcast winter days such a house only requires a minimum of heating, which can be supplied by a small boiler.

Inside the house, all implements are also energy-efficient. The refrigerator and freezer are well insulated, while the washing machine has a low electricity consumption. The light bulbs are of an energy-efficient design, as is the stove with its well-insulated oven.

The way that most of us use electricity to heat our homes is particularly wasteful. At best 40 percent of the fuel required to produce electricity reaches us when we plug it in our wall sockets. The rest is lost in the process of generation and during transmission. In the power station furnaces steam is produced at a temperature of around 1,000°F. In our houses, on the other hand, we normally maintain temperatures of no more than 68°F. Thus when we heat our homes we use electricity, which is produced by *high-grade* heat, to power radiators, which give out only *low-grade* heat. This is a very inefficient use of fuel. And yet the use of electricity for heating has gone up by leaps and bounds in the last forty years.

THE FUEL-HUNGRY HOUSE

Although most modern houses are designed with energy conservation in mind, and are usually well insulated and efficiently heated, older houses can squander large amounts of energy through heat loss. Much of this heat loss can be stemmed by improving insulation. Reducing domestic heat consumption plays a significant part in reducing the pollution caused by burning fuel.

POSITIVE ACTION

Six ways to make an older house more energy-efficient

● **Insulate the attic**
This is one of the simplest and quickest ways of saving energy in the home. In the average house, a 4-inch layer of attic insulation will pay for itself in three years.

● **Fit double-glazing**
Fitting replacement double-glazed windows to an older house is a fairly expensive way of saving energy. Secondary glazing can usually be fitted to existing windows, and will save energy almost as effectively.

● **Insulate cavity walls**
This will save a considerable amount of energy, but the material needs to be chosen with care (see p. 117).

● **Draftproof doors and windows**
Fitting draftproofing throughout a house will usually pay for itself within two years. Care is needed if your house contains any fuel-burning heaters without external flues: these need good ventilation at all times.

● **Insulate tanks and pipes**
Fitting a 4-inch jacket around a hot-water cylinder is the most effective way of preventing household energy loss.

● **Fit radiator thermostats**
Radiators can be made more efficient by installing individual thermostats and fitting foil to reflect heat from external walls behind them.

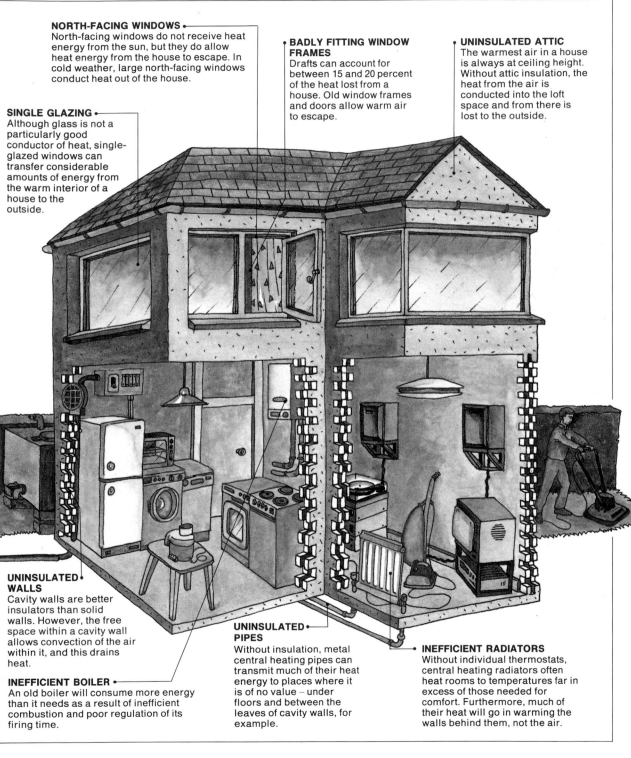

NORTH-FACING WINDOWS
North-facing windows do not receive heat energy from the sun, but they do allow heat energy from the house to escape. In cold weather, large north-facing windows conduct heat out of the house.

BADLY FITTING WINDOW FRAMES
Drafts can account for between 15 and 20 percent of the heat lost from a house. Old window frames and doors allow warm air to escape.

UNINSULATED ATTIC
The warmest air in a house is always at ceiling height. Without attic insulation, the heat from the air is conducted into the loft space and from there is lost to the outside.

SINGLE GLAZING
Although glass is not a particularly good conductor of heat, single-glazed windows can transfer considerable amounts of energy from the warm interior of a house to the outside.

UNINSULATED WALLS
Cavity walls are better insulators than solid walls. However, the free space within a cavity wall allows convection of the air within it, and this drains heat.

INEFFICIENT BOILER
An old boiler will consume more energy than it needs as a result of inefficient combustion and poor regulation of its firing time.

UNINSULATED PIPES
Without insulation, metal central heating pipes can transmit much of their heat energy to places where it is of no value – under floors and between the leaves of cavity walls, for example.

INEFFICIENT RADIATORS
Without individual thermostats, central heating radiators often heat rooms to temperatures far in excess of those needed for comfort. Furthermore, much of their heat will go in warming the walls behind them, not the air.

151

Using off-peak electricity in night storage heaters is convenient and relatively cheap but is still extremely inefficient and has all the environmental problems we have already seen. Only where renewable sources of energy are used for this purpose, such as in Norway where abundant hydro-electricity is available, is the use of electricity for heating purposes really justifiable.

On the other hand, pollution from stoves and household central heating systems is quite small compared with that caused by power stations, industrial emissions, and traffic. Most coal-, gas-, or wood-burning heating systems do not reach such high temperatures. They produce low-grade heat, and (if cleaned coal is used) they do not produce much pollution.

RENEWABLE ENERGY

The all-eggs-in-one-basket school of thought doesn't like renewable energy. Its adherents, among whose number are all the people involved in the traditional methods of producing power, and many of the people who decide how energy is going to be produced, prefer to stick to just one or two ways of power production, no matter how dangerous, dirty, or short-term these are.

But after a decade of research in science departments, by inventors and by commercial companies, there are plenty of other ways of producing power that don't pollute the atmosphere or use up limited resources. Some of these can be used in individual homes, while others are more suited to producing energy on a larger scale, which can then be distributed. As all renewable energy systems depend on the Earth's ultimate energy source, the sun, it makes sense, wherever possible, to use the sun as a direct source of energy.

SOLAR POWER

Solar power is the simplest and least polluting way of supplying energy to the home. The principle behind the flat plate solar collector, for example, is so straightforward that almost anyone could design one. The plate's black surface traps solar radiation and the heat produced is transferred to water. The water can be pumped over the plate, but if you want things to be really simple you can leave it to gravity to do the work. The hot water will rise into a tank, while cold water flows down to replace it. A glass window over the plate acts like a greenhouse pane: it traps the heat so that the temperature of the water rises rapidly. And that's all there is to it.

In places with cold but sunny winter weather, like New Mexico, solar panels have proved to be effective even in the frostiest months. Under these conditions anti-freeze is used in the water that circulates through the solar panels. This flows through a heat exchanger which warms up "clean" water in a secondary system.

In northern countries like Britain, Germany, and Holland, where overcast skies are all too common, solar panels might seem less appropriate. But with all renewable energy sources, the key point is multiplicity. A solar panel might not be able to produce all the heat energy needed on its own, but if used as part of a multiple system of energy sources, it can contribute considerably towards reducing household fuel bills, cutting heating costs by around a third.

By far the best way to make use of solar energy is to make houses into solar heat traps. Passive solar heating systems can be incorporated into existing houses, for example by building conservatories on the south-facing side and by painting the south side a dark color. More sophisticated is the Trombe wall in which air is heated up by the sun shining on a black glass-covered wall. The warm air is circulated through slits in the wall. This system can be used in both old and new houses.

Passive solar heating systems will work with any heat-absorbing surface that can be incorporated into a building. They can be sections of a south-facing wall, piles of black pebbles in a conservatory or even drums filled with water. In one design developed in Israel, a house is heated by capturing the sun's heat in columns made of adobe (sun-dried bricks) which are installed behind a south-facing glass wall and which can then be rotated so that the trapped heat is released into the room.

Houses with passive solar heating systems often cost little more than "ordinary" houses. With high levels of insulation they may be 5 to 10 percent more expensive, but the reduction in heating bills can be between half and three-quarters.

THE SOLAR-HEATED HOUSE

At present, solar energy accounts for a tiny part of total energy use – about one ten-thousandth of that provided by oil. Yet solar energy is easily harnessed in the home. This house shows some of the features that can provide low-cost heat from solar power even in high latitudes.

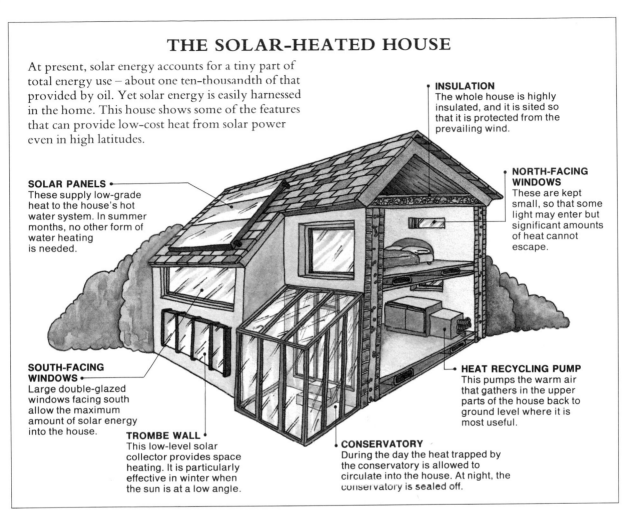

INSULATION
The whole house is highly insulated, and it is sited so that it is protected from the prevailing wind.

NORTH-FACING WINDOWS
These are kept small, so that some light may enter but significant amounts of heat cannot escape.

SOLAR PANELS
These supply low-grade heat to the house's hot water system. In summer months, no other form of water heating is needed.

SOUTH-FACING WINDOWS
Large double-glazed windows facing south allow the maximum amount of solar energy into the house.

TROMBE WALL
This low-level solar collector provides space heating. It is particularly effective in winter when the sun is at a low angle.

CONSERVATORY
During the day the heat trapped by the conservatory is allowed to circulate into the house. At night, the conservatory is sealed off.

HEAT RECYCLING PUMP
This pumps the warm air that gathers in the upper parts of the house back to ground level where it is most useful.

The world's largest solar power plant
In the Mojave Desert in California, the Solar One power plant generates 10 megawatts of electricity with the aid of 1,800 mirrors. This is enough to provide the energy needs of about 2,000 homes. The concentric mirrors reflect the sun's heat to a central boiler where water is converted to steam, which then drives a turbine.

LARGE-SCALE SOLAR POWER

By 1982 the cost of solar-generated electricity had come down sufficiently to make it applicable not only for satellites in space and for electronic calculators but also for powering telephone systems in sunny countries. In Spain remote communities which would be expensive to wire up to the electricity grid are being supplied by solar cells that have been installed locally.

Solar cells – wafers of materials that transform sunlight directly into electricity – are now on the verge of being competitive with conventional sources of electricity. Engineers at Stanford University in California have developed a new type of photovoltaic cell that could convert nearly one-third of the solar energy it receives into electricity, a figure that would have been unthinkable just a few years before. These new cells could soon be used to produce electricity for California's grid in competition with fossil-fueled power stations, and with no pollution whatsoever.

POWER FROM THE WIND

It seems to have taken us a long time to rediscover that the wind is a free source of energy. James Watt's invention of the coal-burning steam engine

Altamont wind farm
This giant Californian wind farm (*above left*) uses windmills of the traditional horizontal axis type. These must face the wind, and can operate only in certain wind speeds.

A Darrieus rotor
This vertically mounted windmill (*above*) will operate in winds from any direction. It has increased stability as gravity does not impede rotation.

led to the windmills of yesteryear falling into disuse; it may be that the problems of burning coal will bring them back.

At one time, the wind was assiduously "gathered" wherever flat ground left its flow unimpeded, or where hills gave access to turbulent air. On the island of Crete there is a marvellous example of wind power: six thousand sets of white sails, turned by the wind that blows across the Lasithi Plateau, work the pumps that irrigate this fertile plain. Designed by Venetian engineers in the 1460s, the windmills are still in daily use.

POSITIVE ACTION

Alternative energy systems for the home

- **Solar water heating**
  In northern temperate latitudes, solar water-heating panels will pay for themselves within five years. The most energy-efficient is the vacuum-tube type: this prevents loss of heat through the back of the collector.

- **Home wind-powered generators**
  Small domestic wind-powered generators are rarely able to provide enough energy for heat or power, but they can be used to charge batteries to provide lighting for outbuildings.

- **Heat recycling pumps**
  Heat recycling pumps can be installed in conventionally heated homes if there is underfloor access. The pump itself consumes relatively little power, while improved heat distribution reduces the total energy needed to keep the house warm.

In these built-up days, unless you live out in the country, it is highly unlikely that wind power could ever play a useful part in providing energy for your home. But that does not mean that we should not be thinking about ways of gathering the energy from the wind and harnessing it for our own purposes. Indeed we would be very foolish to reject out of hand so well-tried and tested a design.

By the end of 1985 over ten thousand windmills were generating electricity in the United States, producing as much electricity as a conventional power station. Ninety-five percent of these windmills are in California, most of them located on giant wind farms. At Altamont Pass several thousand medium-size windmills are grouped together, taking advantage of the strong winds in that area. They produce electricity at 6 to 8 cents per kilowatt, comparable with any large power station.

Wind power is a bargain. In California private individuals can invest in an individual windmill and get paid for the amount of electricity it supplies to the grid. By the end of the century, it is intended that one-fifth of California's electricity will be produced by the power of the wind – so one-fifth less dangerous and polluting conventional electricity will be needed.

In Denmark the world's first offshore wind farm began operating in 1985. A half-mile-long jetty that stretches out into the sea off the Mols peninsula in Jutland supports sixteen windmills which supply over a tenth of the electricity needed by Ebeltoft – a town of 4,000 people. The windmills will pay for themselves in only five or six years. Not far away in North Friesland, West Germany, the local authority has commissioned a wind park for three hundred windmills on the North Sea coast. In many ways solar power and wind power are complementary. Ask any Orcadian and they will tell you cloudy climates are very often windy ones. At Burgar Hill on the Orkney Islands a windmill turbine with a wingspan of 200 feet is being tested on one of the windiest sites in the world. Perhaps wild weather has its advantages after all.

POWER FROM THE SEA

Anyone who has ever been sent sprawling by the sudden slap of a wave will appreciate how much power is captured in that seething mass of salty water. If only a tiny fraction of this power could be converted into electricity, countless fossil-fuel and nuclear power stations would immediately have to shut down their boilers for good.

Wave power is still the province of the enthusiastic inventor. Dozens of nodding, bobbing, pushing, and pulling designs have been tested out, and if any of them had the backing of those in authority, they would soon be put to good use.

On the Norwegian coast at Tostestallen, one machine, a Wells generator, is now in operation and is proving financially viable. It harnesses the power of the waves hitting the coastline through a column of water that rises and falls inside a cylinder. This generates air pressure which drives a turbine. A Norwegian firm is now selling this machine commercially.

Wave power generators do not necessarily have to float on the sea: they can work equally well under it. Not far from Sellafield, Britain's much-criticized nuclear plant, an engineering firm has developed a wave turbine which does just this. It consists of a steel tube which is 260 feet long, weighing 23,000 tons, and is driven into the sea bed. Compressed air is driven through the turbine by

155

oscillating water and this powers a generator. Such generators could be assembled to make up power stations that were entirely under water, well away from the buffeting of the waves, completely nonpolluting and quite invisible from the shore.

Why don't we have wave power stations yet? Because the electricity generated in this way would be marginally more expensive than that produced by conventional means. It would, in return, provide a massive saving in terms of acid rain and radioactive pollution. But extra expense is all that they hear, these people whose immediate interests do not lie in the business of energy conservation or pollution prevention. And that, for them, is the end of the matter.

ENERGY PLANNING FOR THE FUTURE

There is currently just one country in the world – Sweden – where the principles of multiplicity and localism are being applied to energy consumption. The Swedes have drawn up a "sustainable energy policy" which is designed to cut pollution, reduce the reliance on imported oil as a major fuel, and phase out nuclear power altogether.

As recently as 1979 Sweden had the highest *per capita* consumption of imported oil in the world. But in 1981, the Swedish parliament decided on a program that would halve oil consumption by using locally produced fuels instead. They also decided that future energy systems should have the least possible environmental impact.

Following these decisions Sweden started to restructure the way energy was produced and used. Sweden relies on hydroelectric power for generating much of the electricity it consumes. This will not be expanded any further as many people are opposed to damming more rivers on ecological grounds. Fossil fuels still have to be used, but in a more efficient and less polluting way.

A major feature of the energy plan is the development of combined heat and power plants. The city of Stockholm has ordered two coal-fired fluidized bed power stations whose "waste" heat will be used for district heating schemes. These power stations will cause virtually no air pollution. Waste heat from factories and trash incinerators will also be piped into houses and office buildings

instead of more "new" heat being produced quite needlessly.

Reliance on oil for space heating is being cut dramatically by a combination of house insulation program and the use of alternative energy. Geothermal energy, extracted from a layer of water-bearing sandstone 1,600 to 2,600 feet below the surface, is already used as a heat source for houses in the city of Lund. By 1990 at least a third of Lund's district heating supplies will come from below ground. District heating, which was installed over twenty years ago in Sweden, means that homes can be kept warm by whatever source of energy is most sensible in any area.

Solar energy, even this far north, is being taken seriously. Near Uppsala over five hundred detached houses have been constructed whose entire heating needs are met by solar collectors. The collectors gradually heat up water stored in an artificial underground cavern in the summer months, and the hot water can then be circulated through radiators in the winter.

In another sustainable energy scheme, fast-growing willow trees have been planted in an experimental 175-acre plot which will be cut every three to four years as fuel wood.

Obviously Sweden's energy plan cannot automatically be adopted by all other countries. The considerable power of municipal authorities made it easier for Sweden to adopt a policy of flexible and integrated heat and power supply than would be possible in many other places. Nevertheless, Sweden has many lessons to teach to other countries in the merits of multiplicity.

We may have to accept that "clean" power may cost us a little more. We may also have to accept that we must use heat and power more frugally. But we simply cannot afford to foul up the air and the living world around us with the fumes from power stations, factories, and households. This, together with the radioactive waste that is piling up in storage tanks at power stations, that is being discharged into the oceans and that is blowing about in the wind, is *our* waste, and it is up to us to put a stop to it.

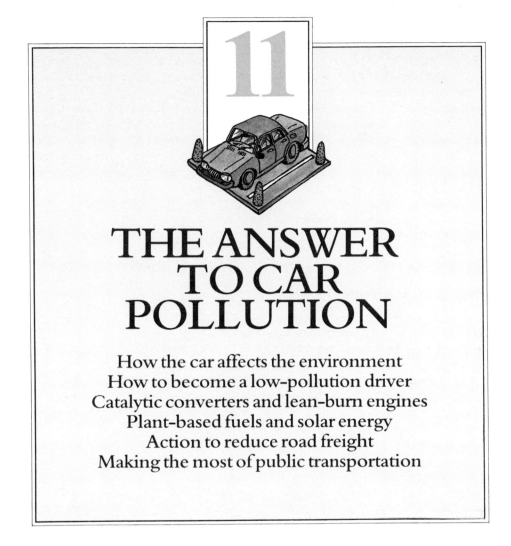

THE ANSWER TO CAR POLLUTION

How the car affects the environment
How to become a low-pollution driver
Catalytic converters and lean-burn engines
Plant-based fuels and solar energy
Action to reduce road freight
Making the most of public transportation

About 30 million cars are made every year. Each one will offer its owner complete freedom of movement. In exchange it will get through about three tons of gasoline, will grossly pollute the atmosphere, will kill wildlife and perhaps pedestrians, and will offer its driver a chance to kill him or herself into the bargain. It may not sound like a very good bargain, but it is the one that has been struck.

In the days of horse-drawn transportation, every main street was as much a playground and a place for gossip as a thoroughfare for travelers. All the necessities of daily life were close at hand, and so there simply wasn't the need for much getting about. But all that has changed. For most of us, no amount of persuasion, no amount of cogent argument, nor any amount of nostalgia will ever part us from our cars, because our world is now structured around rapid transit. Despise the car though we may, we have come to depend on it. We think nothing of traveling a hundred miles to visit friends. We take it for granted that, through the car, our homes and workplaces are far apart, and that the supermarkets where we shop cannot be reached on foot. We even accept (reluctantly) that the car is a social yardstick by which we will be judged.

As a means of travel, the car is undoubtedly a mixed blessing. We have to live with it, but at the same time we have to do a lot more to bring the car's worst effects under control.

WHATEVER HAPPENED TO THE "FUEL CRISIS"?

The man who founded OPEC, the Venezuelan Perez Alfonso, did so because he was so concerned about the speed with which oil was being used up. He considered the car to be a curse: whenever possible, he traveled by bicycle. He did have a car, but it was not some chrome–laden monster. Instead he owned an elderly vehicle that was permanently parked in his garden and covered with vegetation.

He understood the importance of fuel conservation.

In the seventies, there was a lot of talk about world oil reserves, how they could be conserved, and how much longer they might hold out. But gradually it became an unfashionable subject – people got bored with it. Nowadays you might be forgiven for thinking that the Jeremiahs of the seventies had gotten it wrong, because here we are with plenty of gasoline available for our cars, and instead of costing more, it actually costs less.

THE PRICE OF MOTORING
How the use of cars affects the environment

The 500 million motor vehicles on the world's roads have adverse effects on all the realms of the natural world – land, air, and, through oil pollution, the sea. Because many of these effects have crept up gradually, we now accept them. But if the car was a new invention, we might well feel that this price was too great.

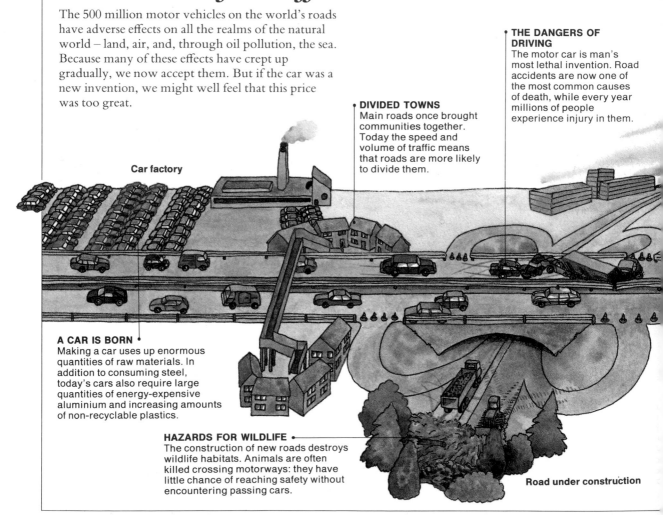

THE DANGERS OF DRIVING
The motor car is man's most lethal invention. Road accidents are now one of the most common causes of death, while every year millions of people experience injury in them.

DIVIDED TOWNS
Main roads once brought communities together. Today the speed and volume of traffic means that roads are more likely to divide them.

Car factory

A CAR IS BORN
Making a car uses up enormous quantities of raw materials. In addition to consuming steel, today's cars also require large quantities of energy-expensive aluminium and increasing amounts of non-recyclable plastics.

HAZARDS FOR WILDLIFE
The construction of new roads destroys wildlife habitats. Animals are often killed crossing motorways: they have little chance of reaching safety without encountering passing cars.

Road under construction

However, we are currently living in a fool's paradise (if paradise is the word for a world that is choking on car exhausts). Although the price of oil may go up and down, world oil reserves go only in one direction – down. If the price of a barrel of oil falls by half or two-thirds, it doesn't mean that there is suddenly any more of it to go around. On the contrary, when oil is cheap, consumption gradually goes up, and the reserves go down that much faster. It doesn't take a degree in economics to see that

eventually, despite all the seesawing, the price of oil will rocket as the reserves plummet.

But we continue to use oil as if there were no tomorrow. Total world consumption amounts to about 2.8 billion tons. Half of this is used in transport as a whole, while about a third goes into the internal combustion engines of cars and trucks.

The world's oil deposits have accumulated in the Earth's crust over hundreds of millions of years – but at current rates of consumption they will have

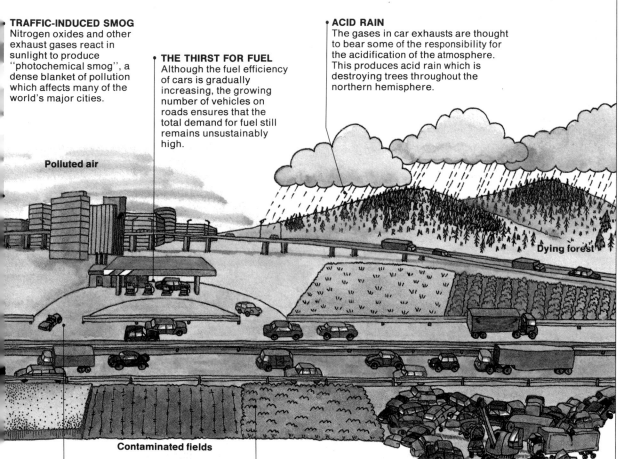

TRAFFIC-INDUCED SMOG
Nitrogen oxides and other exhaust gases react in sunlight to produce "photochemical smog", a dense blanket of pollution which affects many of the world's major cities.

THE THIRST FOR FUEL
Although the fuel efficiency of cars is gradually increasing, the growing number of vehicles on roads ensures that the total demand for fuel still remains unsustainably high.

ACID RAIN
The gases in car exhausts are thought to bear some of the responsibility for the acidification of the atmosphere. This produces acid rain which is destroying trees throughout the northern hemisphere.

Polluted air

Dying forest

Contaminated fields

AND LOSS
here are estimated to be bout 7 million miles of urfaced roads in the Western industrialized ountries. All the land they over is permanently lost o agriculture and wildlife.

CORRIDORS OF LEAD
In countries where leaded gasoline is used, exhaust fumes create "lead corridors" alongside main roads. The ground in these corridors can be contaminated with lead to a dangerous extent.

THE END OF THE ROAD
Car manufacturers have little interest in building cars that last. Changing fashions and rapid deterioration ensure that the average life of a car is far shorter than it could otherwise be.

Breaker's yard

been used up in just a few decades. Much depends on the rise or fall in worldwide demand, oil extraction techniques and new oil finds, but most oilmen agree that all the major deposits around the world had *already* been found by 1980, and that all the recoverable oil reserves will have been used up in about thirty years' time. By the time all today's children are parents, most of it will have gone.

You might ask what is the point of going to great lengths to conserve fuel if it is going to run out anyway. What difference will a few years make? Well, using fuel lavishly doesn't just mean that we will have to give up gasoline engines that much sooner. It also means that the gasoline engine's unpleasant side effects will become even more serious, because fuel will be turned into pollution at an even faster rate. Saving fuel means spreading out and reducing these effects – and giving us a chance to find some sensible alternatives to the internal combustion engine before the oil really does run out.

THE CAR POPULATION

In the industrialized countries, the numbers of cars are rapidly catching up with the human population. This table shows how human and car populations compared in 1983.

	Number of cars	People per car
United States	126,000,000	1.75
Canada	10,800,000	2.13
New Zealand	1,401,000	2.14
Australia	6,600,000	2.18
Germany	24,689,000	2.47
Switzerland	2,521,000	2.51
France	20,600,000	2.58
Sweden	3,007,000	2.76
Italy	20,000,000	2.86
Norway	1,383,000	2.90
Netherlands	4,770,000	2.94
Belgium	3,263,000	3.02
Finland	1,411,000	3.33
Great Britain	15,854,000	3.34
Denmark	1,390,000	3.60
Spain	8,714,000	4.25
Japan	26,386,000	4.40

THE LUST FOR SPEED

For many people, the series of oil price "shocks" which started in the early seventies were an unexpected lesson in frugality. Through some very modest scrimping and saving, annual oil consumption throughout the Western world dropped by about one-tenth between 1973 and 1983.

Much of this was achieved through changes on the roads. Many countries imposed national speed limits for the first time. In the United States, where the presence of rapidly increasing oil imports was beginning to be felt on the balance of payments, a national speed limit was introduced. Furthermore, some steps were taken to make cars more sparing with fuel. "Gas guzzlers" went out of fashion, and by 1985 the standard of fuel efficiency in the United States had doubled, while on the other side of the Atlantic the figures were even better.

But one lesson that did not seem to sink in was the connection between speed, waste, and pollution. It is one thing to tell people to drive more slowly, it is quite another thing to get them to comply. Many cars are made with engines whose power vastly exceeds the needs of the motorist. As a result, our roads have become proving grounds for cars that are designed to break speed limits and aggravate pollution. The psychological addiction to speed has still not diminished: too many people are still in love with the fast car.

SLOWING DOWN TO REDUCE POLLUTION

There are now 350 million cars in use worldwide and together they produce about 13,000 billion cubic yards of exhaust fumes every year. It has been calculated that the fumes from West Germany's 25 million motor vehicles alone would cover the entire country with a layer of poisonous fumes $6\frac{1}{2}$ feet high. This would suffocate all life, if it were not for the fact that much of it is carried up into the atmosphere.

If you drive at high speed the output of exhaust gas greatly increases. Nitrogen oxides, which are suspected culprits in tree damage, are produced at twice the rate in a given distance from a car traveling at 100 mph compared with one traveling at 55 mph. The output of poisons such as carbon monoxide and hydrocarbons also rises sharply and so does gasoline consumption.

In addition there is the risk to personal health. Where there are no speed restrictions, accident statistics read like a medical bulletin during an epidemic. In Germany there was a total of 179,000 car crashes in 1984 in which over 10,000 people died. Of the whole population, one person in a hundred was injured within that twelve months.

Speed limits on highways – although often broken – reduce the figures dramatically. In 1974 a 55 mph speed limit was introduced in the United States. There were over 55,000 road fatalities the preceding year when the highway speed limit was 65 mph. Ten years later this figure had dropped to 44,000 despite a doubling of the traffic volume.

Reduced speed is the simplest way to reduce pollution, conserve fuel, and survive to drive another day.

WHY WE DON'T NEED MORE ROADS

There is a law that applies to building roads, and it is this: no matter how much land you bury under tarmac to relieve traffic congestion, enough cars will be found to clutter it up.

New roads destroy our natural heritage. Highways are the worst example. These great stretches of concrete ravage the countryside, cutting across fields, knocking down woodland, bulldozing away hills and slicing in half villages and small towns.

To the planner, the rights of people in the path of a new road are the lowest of priorities. In order to shave a few minutes off a journey they condemn households to live with the constant rush of traffic, or force people to leave areas where they were born and brought up. Nor are these few saved minutes worth the waste of animal life. Hundreds of thousands of birds and small mammals are slaughtered needlessly every day by the cars that go crashing through our countryside. The toll is even worse with new roads because they often cut across traditional animal pathways.

Many countries have already sacrificed far too much good land to the car. There should be no such thing as new roads.

LEAD-FREE GASOLINE

There is no debate about whether or not lead is good for you. It isn't. The Romans used to inadvertently poison themselves by steeping wine in lead cauldrons, and in the days when plumbing involved the use of lead piping, the threat of poisoning was a recognized hazard. Drawing off the water that had stood in the pipes overnight was an essential part of the daily domestic routine.

The tarmac jungle
In our car-dominated society, the motorway has precedence over every other form of land use. In order that motorists should not have to suffer any break in their journeys by lining up at junctions, huge areas of land are given over to complex interchanges. The air pollution these create makes the land around them dangerous for cultivation, while constant traffic noise makes the area unfit for habitation.

Dissolved lead has since given way to airborne lead. Gasoline fumes are full of it, and in this way the lead is distributed far and wide. Lead concentrations even in unpolluted Greenland have risen by between five hundred and one thousand times since prehistoric times, and most of this has happened in the last hundred years. Yet there is a quite straightforward solution in the form of lead-free gasoline.

Lead is highly dangerous to children because it can damage brain development. Some countries have recognized the importance of reducing lead in the atmosphere and are ensuring that it is widely available if not obligatory. But for other countries, it appears to be too expensive. But lead-free gasoline doesn't have to be costly. Governments are quite adept at subsidizing such nonsenses as food mountains: why can't they subsidize something sensible like lead-free gasoline?

This is precisely what has happened in West Germany. Public opinion eventually forced the government to bring the price of lead-free gasoline down to below the price of ordinary gasoline. From 1986 all new cars sold in West Germany had to be able to run on lead-free gasoline. By the following April one-third of the gasoline sold in the country was lead-free.

The oil companies and car manufacturers and the people with shares in their businesses may not like making these changes. But compared to the expense in terms of children's health and pollution of the land the cost of implementing lead-free gasoline and the cars to run on it is negligible. There is no reason why anyone should have to drive a car that pours out this known poison.

CARS AND TREES

No one knows exactly how much guilt the car bears for acid rain, and how much is due to power production (see p. 147), but the precise figures are irrelevant. Car exhausts kill trees, and that much is certain.

The destruction of forests isn't just a matter of picturesque views being spoiled, or wildlife losing its home. In many places, trees hold together the very land itself so that we can drive thousands of polluting cars and trucks across it.

THE HIGH-POLLUTION DRIVER

All gasoline-powered cars harm the environment, but the polluting effects of driving can be minimized by taking care in the way a car is used. Those who do not take care simply aggravate pollution and further, reduce the lives of their cars, and sometimes their own lives as well. Although high-pollution drivers are often seen in old vehicles, the speed-obsessed owners of sports cars are equally guilty of environmental thoughtlessness.

POSITIVE ACTION
How to avoid being a high-pollution driver

● **Choose your car carefully**
When buying a car, put pollution control factors high on your list of desirable characteristics. The engine should have at least one of the features in the engine illustrated on p. 165.

● **Check your driving technique**
Driving smoothly creates less pollution. Avoid sudden acceleration, as this greatly increases the car's output of exhaust pollutants. Frequent hard braking wastes the energy that gasoline provides.

● **Keep your car properly maintained**
Preventing rust is better than treating it with toxic rust-removers. Keep the car free of dirt and, especially during the winter, road salt. If you have a garage, use it.

● **Have the engine tuned regularly**
Make sure that the engine is regularly tuned. This will ensure that it burns gasoline with the maximum efficiency.

● **Avoid the car-change habit**
Do this as little as you have to. Looking after a car is far better for the environment than exchanging it for a new one.

● **Leave your car at home**
Try not to fall into the habit of using the car when walking or cycling would be better for you, or when using public transportation would be quicker.

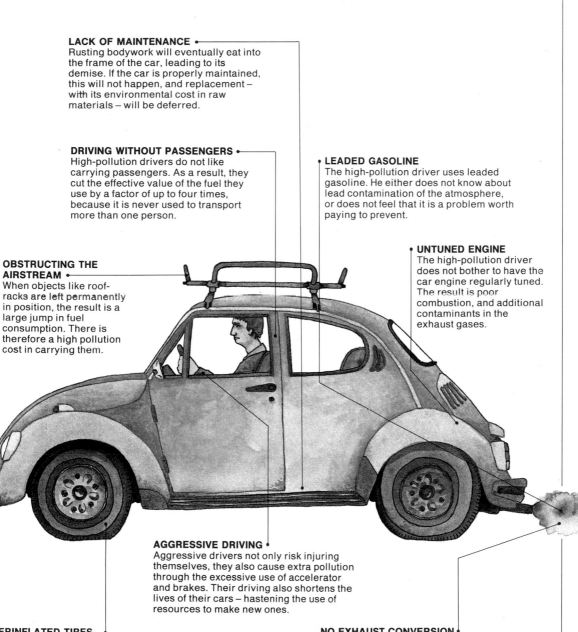

LACK OF MAINTENANCE
Rusting bodywork will eventually eat into the frame of the car, leading to its demise. If the car is properly maintained, this will not happen, and replacement – with its environmental cost in raw materials – will be deferred.

DRIVING WITHOUT PASSENGERS
High-pollution drivers do not like carrying passengers. As a result, they cut the effective value of the fuel they use by a factor of up to four times, because it is never used to transport more than one person.

LEADED GASOLINE
The high-pollution driver uses leaded gasoline. He either does not know about lead contamination of the atmosphere, or does not feel that it is a problem worth paying to prevent.

OBSTRUCTING THE AIRSTREAM
When objects like roof-racks are left permanently in position, the result is a large jump in fuel consumption. There is therefore a high pollution cost in carrying them.

UNTUNED ENGINE
The high-pollution driver does not bother to have the car engine regularly tuned. The result is poor combustion, and additional contaminants in the exhaust gases.

AGGRESSIVE DRIVING
Aggressive drivers not only risk injuring themselves, they also cause extra pollution through the excessive use of accelerator and brakes. Their driving also shortens the lives of their cars – hastening the use of resources to make new ones.

UNDERINFLATED TIRES
Driving with underinflated tires increases gasoline consumption. This means that more gasoline is needed to cover a given distance, and therefore more exhaust pollution will be created.

NO EXHAUST CONVERSION
Without a catalytic converter, the car's exhaust system pours nitrogen oxides, carbon monoxide, and sulfur dioxide into the atmosphere. These poisonous gases add to the pollution of the atmosphere.

THE ANSWER TO CAR POLLUTION

The Swiss are as excessive in their use of cars as any rich nation. They also have huge numbers of visitors who come to ski on their snow-covered mountains in the winter and to walk and climb in the high Alps in the summer. A great many of these tourists come by car. In the last few years forests in Switzerland have begun to show signs of very serious damage. There is no doubt that the cars of both the tourists and the local people are contributing greatly to this calamity.

Although the forests that cover the mountain slopes of Switzerland are a delight to look at, they also perform a crucial task. They protect the villages, farms, and roads in the valleys below from avalanches and landslides. It has been estimated that it would cost the country billions of dollars to build barriers along roads and around villages if the "free" protective function of the mountain forests were to fail. And as a result of forest die-back that is fast becoming a possibility.

The town of Bristen in the mountains south of Lucerne is typical of many towns and villages whose future has been cast into doubt. The forests on the slopes above the town are sick. Trees weakened by air pollution are becoming infested with bark beetles and are having to be cut down. As the trees are felled the danger of avalanches and landslides becomes even greater. Heavy snowfalls are no longer absorbed by the tree canopy and snow and soil debris can tumble down the mountainsides unimpeded by the forest which once stood in their way. In the summer of 1985 the people of Bristen built several huge soil terraces above the town to intercept avalanches, but nobody knows if this will provide adequate protection. They have planted young trees as well, but it is uncertain whether they will survive in the polluted air. To make matters worse, the Swiss Forestry Office estimates that one-tenth of the mountain forests will go the same way in the next few years. This will put at least 150,000 people at risk.

But you don't have to live in the Alps to suffer from tree damage. Even in lowland areas, tree roots form a lattice which holds the soil in place. When the trees die, the soil is washed away and the result is erosion – and all this is triggered by the quite unnecessary pollution from car exhausts.

Urban smog
Blankets of exhaust fumes (*above*) in many of the world's cities make the air irritating to humans and toxic to plants.

The airborne defoliant
In the majority of the world's cars, exhaust fumes (*below*) leave the engine without any form of treatment.

CUTTING POLLUTION WITH THE CATALYTIC CONVERTER

There is a perfectly simple solution to a large part of the car pollution problem, and this is the catalytic converter.

A catalytic converter, built into the exhaust system of a car, forces exhaust gases to react with each other. The result is that the harmful constituents are almost completely changed into safe ones. As the exhaust gases are blown through the converter's honeycomb-like grille, the platinum catalyst prompts reactions between the gases which would otherwise not take place. The converter has no moving parts, and, once manufactured, does not require any additional energy.

Catalytic converters were first developed and tested in extensive road trials in the early seventies with Europe leading the field. However, when the 1973 oil crisis caused fuel prices to leap upward, European governments and motor companies were reluctant to fit cars with catalytic converters because it would have been an additional financial burden on the car owner.

In America there were no such inhibitions. In an attempt to reduce air pollution levels in cities, new exhaust regulations in 1975 required car manufacturers to ensure drastic reductions of emissions of nitrogen oxides, carbon monoxide and hydrocarbons from all new cars. As a result, American exhaust regulations became the strictest in the world – and strictest of all in California, where the average nitrogen oxides output is now twelve times lower than from European cars.

WHY AREN'T CONVERTERS USED EVERYWHERE?

If catalytic converters are such a good thing, why don't all countries insist on them? The problem, as with lead-free gasoline, comes down to money – how much of it people are willing to pay to prevent pollution. A catalytic converter adds about 5 percent to the cost of a car. It also increases fuel consumption by about that figure (although of course this can be offset by driving more slowly), and it requires lead-free gasoline and a slightly modified engine design. It seems that for some countries this is too much for preserving forests, particularly if *their* forests aren't the worst affected.

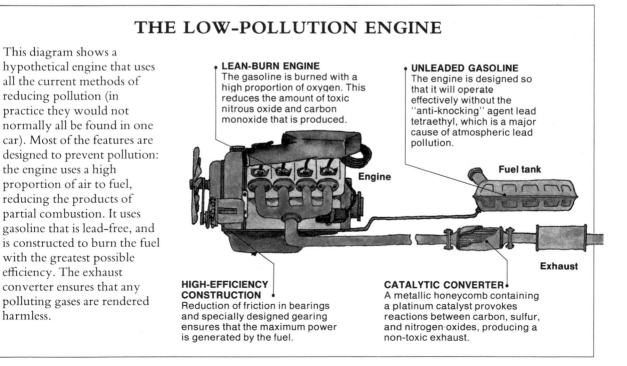

THE LOW-POLLUTION ENGINE

This diagram shows a hypothetical engine that uses all the current methods of reducing pollution (in practice they would not normally all be found in one car). Most of the features are designed to prevent pollution: the engine uses a high proportion of air to fuel, reducing the products of partial combustion. It uses gasoline that is lead-free, and is constructed to burn the fuel with the greatest possible efficiency. The exhaust converter ensures that any polluting gases are rendered harmless.

LEAN-BURN ENGINE
The gasoline is burned with a high proportion of oxygen. This reduces the amount of toxic nitrous oxide and carbon monoxide that is produced.

UNLEADED GASOLINE
The engine is designed so that it will operate effectively without the "anti-knocking" agent lead tetraethyl, which is a major cause of atmospheric lead pollution.

Engine

Fuel tank

Exhaust

HIGH-EFFICIENCY CONSTRUCTION
Reduction of friction in bearings and specially designed gearing ensures that the maximum power is generated by the fuel.

CATALYTIC CONVERTER
A metallic honeycomb containing a platinum catalyst provokes reactions between carbon, sulfur, and nitrogen oxides, producing a non-toxic exhaust.

THE ANSWER TO CAR POLLUTION

The decisions about what kind of pollution your car will produce are taken by car-makers, oil companies, and governments. Some car-makers have been less than enthusiastic about catalytic converters because they think that such devices might be superseded by more modern engine technology. One such innovation is the "lean-burn" engine which mixes more air with its fuel, producing fewer harmful exhaust gases. But more often, commercial interests ensure that keeping up turnover is placed before reducing pollution.

The only way this can be reversed is for motorists to make it clear that they are willing to pay the extra. If you buy a car without a converter, every journey it makes will be adding to the destruction of the natural world around you – something that surely cannot be valued in dollars and cents?

ARE DIESEL ENGINES LESS POLLUTING?

The diesel engine is enjoying a growth of popularity with motorists because of its economic use of fuel. On the face of it, this sounds beneficial – after all, fuel conservation is a sound aim.

In a diesel engine, less of the fuel's potential energy is wastefully converted to heat, and more is converted to power. This is because the fuel is compressed more tightly than in a gasoline engine, so that when it does explode, it does so very thoroughly. The diesel engine also has the advantage of needing fewer parts, and consequently it is longer-lasting and less likely to need maintenance. Again, from a conservationist's viewpoint, this is a good thing: the longer-lasting cars are, the less quickly they need to be turned out of car factories and on to the streets.

Unfortunately, the diesel has a number of severe disadvantages. It can give off clouds of soot particles, and these are suspected of causing cancer if inhaled regularly in high concentrations. Furthermore diesel engines fitted to trucks are usually constructed quite differently than those used in passenger cars. To achieve maximum engine power the fuel is injected directly into the engine combustion chambers and this greatly increases the emission of harmful fumes and soot, as anyone who has stood behind a straining diesel-powered truck will know.

On balance, diesel *cars* are probably less polluting than their gasoline-driven counterparts. However, diesel *trucks* are a different matter. The exhaust fumes they produce are responsible for a large part of the pollution caused by traffic.

THE GROWTH OF TRAFFIC

In the ten years from 1970 to 1980, the number of motor vehicles in the world (private cars and goods vehicles) almost doubled. This fact alone is enough to explain the sudden rise in atmospheric pollution problems that the world is suffering from. Since then, despite the recession, the increase in vehicle numbers has shown no signs of reducing. This inexorable rise is the result of two factors. The first is the continuing spread of car ownership, and the second is the increase in road freight transport.

Year	Vehicle numbers
1970	245 million
1975	327 million
1980	422 million
1985	520 million
2000	800 million?

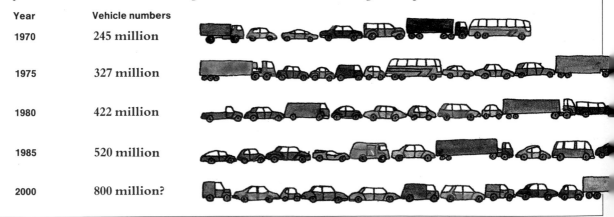

DO WE NEED THE JUGGERNAUT?

The truck is almost as polluting and damaging to our civilization as the car. Anybody can see, without recourse to a mass of tedious figures, that the volume of truck traffic on the roads of every western country is enormous. Much of this traffic is completely unnecessary. On highways, identical products move in opposite directions, as businesses rush to meet orders at opposite ends of the country.

Do we really need this much transportation of goods? The answer is no. A vast amount of road transport traffic – probably at least three-quarters – could be made redundant if the principle of localism were adhered to, with merchandise being manufactured on a smaller scale and nearer to the places where it was to be sold.

Many people will immediately object that this would throw transportation workers – the truck drivers – into unemployment. Well, they would not become unemployed. Localism would create more work than it would lose. A multitude of small factories will create more employment than a few huge ones (and they would be far more pleasant to work in too). The great task of making regions more self-sufficient – as they used to be before the truck was invented – would absorb a vast quantity of labor. Machines would be thrown out of work, not people.

Although by far and away the most polluting mode of transportation of all, the truck is extremely convenient for businessmen. It carries goods from door to door. It is cheap because it is heavily subsidized. But if we really must have truck transportation, we should keep it local. There should be a system of containers, along the lines of what already exists in shipping, that could be transported over most of their distance by rail and

water, and only picked up and delivered at each end of the journey by road transport.

It should be remembered that for most regular goods transportation speed is irrelevant. Provided the producer is able to feed his production into the pipeline of the transport system as he requires to, and the consumer is able to have a constant supply at his end of it, it doesn't matter that much how long each batch of goods takes to do the journey. It is the continuity which is important.

FUEL FROM PLANTS

It is an unfortunate fact that in picking oil to power our cars, we have saddled ourselves with a fuel that is full of substances that pollute the air when they

---

POSITIVE ACTION

**Everyday action
to reduce road freight**

- **Buy local products**
  Avoid buying products that have been carried long distances if you can obtain local alternatives.

- **Object to "improvements"**
  Many road improvement schemes are designed to facilitate access by heavy trucks. Objecting to these schemes may prevent trucks spreading pollution further along the road system.

- **Keep trucks and people apart**
  As major roads become more and more congested, trucks spread into residential areas in search of less busy routes. If you see this happening, complain to the company involved.

- **Reduce empty loads**
  Large amounts of fuel are wasted by trucks returning empty to their point of origin. Materials for recycling (see p. 91) can always provide useful return loads.

are burned. But there are plenty of fuels that do not contain unwanted chemicals. Some of them can be produced from plants. These are known as "biomass" fuels – fuels that are created by fermentation of a crop, the fuel usually being an alcohol.

When oil prices went up so dramatically in the seventies, the Brazilian government decided to encourage the use of home-produced fuels. The production of ethanol (the alcohol that gives drinks their effect) from sugar cane, cassava, sorghum, sweet potatoes, maize, or wood is now well established. The huge area of land allows the Brazilians to take full advantage of the potential of making fuel from agricultural crops. The country now produces over 1 billion gallons of the stuff, and the most efficient distillation processes can produce $18\frac{1}{2}$ gallons of alcohol from each ton of cane. About half a million cars are being run on pure alcohol. Some of it is even being exported.

The beauty of running a car on alcohol is that you can practically breathe the car's exhaust. Pure alcohol contains only carbon, oxygen, and hydrogen, and burns to produce carbon dioxide and water vapor – just what we ourselves exhale.

It doesn't need a Brazilian climate to produce alcohol fuel. Sweden is experimenting with distilling it from fast-growing willow trees, and in America, corn that is normally grown as cattle feed has been used to produce alcohol for cars. Most of the countries that have introduced ethanol have decided on a mixture of ethanol and gasoline. This saves on gasoline while making the addition of lead unnecessary, and still cuts down substantially on gasoline-based pollution.

Another alcoholic alternative fuel is methanol, or wood alcohol, the bane of the home distiller. This can be made from lignite coal, wood, natural gas, or from methane produced in digester plants (see p. 30). At present it is more expensive to produce than gasoline, but processes are being developed which could make it very much more competitive in price in the future. Methanol, too, is less harmful to the environment than gasoline, with which it can be blended. Engines running on methanol require considerably more fuel, but it is ideal in lean-burn engines, which mix it more thoroughly with air. The conversion of natural gas into methanol is particularly promising, and a far better way to use it than turning it into fertilizer.

Natural gas can itself be used as a fuel for cars and trucks. In Holland it has long been a very popular fuel, partly because it is cheaper than gasoline, partly because it is very much cleaner. A mixture of butane and propane is recovered from natural gas

Filling up with alcohol In this Brazilian service station, cars fill up not with gasoline but with alcohol. Alcohol has two advantages as a fuel. Because it is produced from plants, its supply is not finite, like that of oil. Furthermore, compared with oil, it is chemically very pure, and burns without producing a mixture of polluting gases. Experiments with plants in northern countries suggest that alcohol could be a useful fuel or fuel additive in nontropical areas also.

and also gas that has been pumped out of oil wells. Natural gas can also be used in diesel engines – they burn very much more cleanly when running on a combination of diesel oil and natural gas than on diesel oil alone.

Diesel engines can also be run on vegetable oils such as sunflower, soya, and olive oil. Enough oil can be obtained from the sunflower crop of one field to fuel a tractor while it plows ten fields of similar size. Brazil is planning to obtain nearly one-fifth of its diesel oil from plants.

All these plant fuels have a number of teething problems. Ethanol, for example, requires vast areas of land to produce realistic quantities, while at present methanol is extremely expensive. They may not be the complete answer to the problem of gasoline pollution, but they are useful stepping-stones towards it.

CARS AND THE GREENHOUSE EFFECT

Before the Industrial Revolution, fossil fuels such as coal, oil, and gas were locked up within the Earth's crust. Virtually all the carbon dioxide that was released by man came from wood fires. Now carbon dioxide is being poured into the atmosphere from a whole range of sources, including forest clearance, open fires, stoves, central heating, power station burners, and of course cars and trucks.

Although it makes up only a tiny fraction of the Earth's atmosphere, carbon dioxide is crucially important because it regulates the amount of heat that the Earth absorbs from the sun. The more carbon dioxide there is, the better an insulator the atmosphere becomes, because it reflects heat that would normally escape from the Earth to space. As carbon dioxide levels rise, a growing proportion of the solar radiation is becoming trapped inside the Earth's atmosphere and is being reflected back to the Earth's surface. This is what is meant by the "greenhouse effect" – and we are the rapidly ripening tomatoes.

There are about as many opinions on the results of the greenhouse effect as there are scientists studying it. The "traditional" view is that the Earth will get warmer and warmer, but other scientists predict that the results will be much more complex, and that some parts of the world might even get colder.

But one thing is sure, any sudden change to the climate of our planet is bound to work against us. This isn't just pessimism. We have settled our planet in accordance with the conditions as they prevail at present. If these change, we will find that all our established patterns of agriculture and population distribution will have to change.

If the greenhouse effect really gathers momentum, the expected changes in temperature will not be distributed evenly across the planet's surface. The increase in the polar regions might be highest – perhaps as much as 11–14°F. This would almost certainly lead to melting of parts of the Antarctic ice sheet. This could make life very uncomfortable for the inhabitants of such major cities as London, New York, Peking, and Amsterdam. These would all be flooded. Huge areas of low-lying, fertile farmlands would be lost.

What can we possibly do about such a global problem? The answer is to stop burning carbon when we don't really need to – for example, in fuels for our cars.

RENEWABLE FUELS

If we carry on wasting carbon-based fuels in power stations, in badly insulated houses and in energy-inefficient motor vehicles, we will have doubled the natural carbon dioxide content of the atmosphere by 2050. Putting all our faith in biomass fuels won't help, because they only recycle this extra carbon dioxide. We need to cut out carbon-based fuels wherever possible, and instead use the pure energy available from the sun.

The first ever grand prix race for solar-powered vehicles took place in Switzerland in 1986. Nearly a hundred cars took part, and each one covered nearly 250 miles in six stages. The fastest car, a two-seater studded with solar cells, reached an average speed of 30 mph. Its top speed was twice that. Most of the vehicles were equipped with batteries which could power their electric motors in cloudy conditions.

Now this race may well sound like a somewhat eccentric way of passing a few hours, but its message is quite serious. Solar power works, and cars that run on renewable energy sources must eventually replace the gasoline-driven car.

Solar travel
This experimental solar car from Switzerland (*left*) is driven by light-collecting cells mounted on its hood.

Electric transport
The development of electric vehicles (*below*) has been centered around the search for lightweight batteries that can store large amounts of energy. If powered by "renewable" electricity, they would be genuinely nonpolluting.

Obviously, solar cells don't work in cloudy climates, but there is a renewable fuel that may one day be created by sunlight. This is hydrogen, immeasurable quantities of which are available in water. Hydrogen can be burned with oxygen to create a source of power – and the only substance to come out of the exhaust pipe is steam! Several experimental vehicles running on hydrogen have been used in Berlin since 1984. In the United States too, car companies have been testing vehicles that can run on hydrogen. Although it is at present a laboratory technique, it should eventually be possible to use "photolysis", which can produce hydrogen from water using solar energy.

REVIVING PUBLIC TRANSPORTATION

Renewable fuels are still a long way in the future. Until they become fully practical, we need to use the resources we do have a little more wisely.

The battle between the private motor car and public transportation has been raging for years now. Wherever the motor car is on the offensive, the railways, buses, and trolleys are on the retreat. Yet compared with private cars, all forms of public transportation are vastly more energy-efficient and vastly less destructive to the environment.

In times gone by, before the mass ownership of the motor car, public transportation was an efficient and much used way of getting around. The car has

TRANSPORT AND FUEL EFFICIENCY

Public transportation is exceptionally fuel-efficient. Here the total amount of urban "passenger travel" – the distance traveled on a gallon of fuel multiplied by the number of passengers carried – is compared for six different ways of traveling.

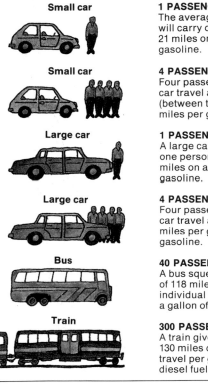

Small car — **1 PASSENGER**
The average small car will carry one person 21 miles on a gallon of gasoline.

Small car — **4 PASSENGERS**
Four passengers in the car travel a total (between them) of 84 miles per gallon.

Large car — **1 PASSENGER**
A large car will carry one person only 14 miles on a gallon of gasoline.

Large car — **4 PASSENGERS**
Four passengers in the car travel a total of 56 miles per gallon of gasoline.

Bus — **40 PASSENGERS**
A bus squeezes a total of 118 miles of individual travel out of a gallon of diesel fuel.

Train — **300 PASSENGERS**
A train gives a total of 130 miles of individual travel per gallon of diesel fuel.

economical in its use of space, and it ensures a smaller volume of traffic uses that space. In countries where the public transportation systems are taken seriously, it is also far, far cheaper, both for the passengers and for the environment.

Public transportation really comes into its own in cities, where there are large numbers of people wanting to travel every day but little space for them to move about in. But to make public transportation work, we need to rid ourselves of the idea that traveling in your own car is somehow a superior way of getting about. It isn't. Having a whole car to yourself is about the most antisocial method of traveling you can possibly devise.

HEALTH HAZARDS FOR PASSIVE DRIVERS

Nobody wants to become ill through being a passive smoker, so why should anyone have to accept becoming ill as a "passive driver"? The pollution caused by cars is quite as intense as cigarette smoke in many city streets, but too few drivers think of the effect that *their* cars have on other people.

Since the car became available to the ordinary person, it has acquired a right of precedence which until recently nobody has dared to challenge. Everything makes way for the car, and the right of drivers to make life unbearable for everybody else is simply accepted. The tables need to be turned: anyone who drives in an area with adequate public transportation should need an unimpeachable reason for doing so.

What would happen if all major cities gave priority to public transportation? In 1986 Friends of the Earth launched a campaign in London called "Cities for People" which set out the advantages. More public transportation would reduce the amount of noise, congestion, and pollution. It would reduce the number of road accidents. It would help to maintain areas of public open land and private housing. It would clear the roads of unnecessary vehicles, making them easier and safer for use by essential vehicles such as police cars and ambulances. Improved transportation systems would make life much easier for the people who cannot or do not wish to drive in the first place.

All it needs is a little imagination and a little

since undermined these more sensible methods of travel, but in doing so, it may well have become a victim of its own popularity. In many cities, the average journey speed of the car driver is steadily going down, not up. Car drivers spend more and more time in stationary lines, enjoying the so-called convenience of car travel, surrounded by thousands of their kind, each of whom is pumping exhaust fumes into the air from a uselessly idling engine. It is all a far cry from the kind of driving that is shown in car advertisements.

Public transportation, although it does not do away with pollution or the need for roads, minimizes the evil of both. It is much more legislation: tax disincentives for owners of more than one car; buses and trains running throughout the day and night at regular intervals and designed to link up with theaters, restaurants, football stadiums; transportation networks that connect urban and rural areas – all the things that the car-maker will throw his arms up at.

LAST BUT NOT LEAST: THE BICYCLE

Bicycles are the most economical and pollution-free form of transportation yet invented. They run on nothing more than human energy derived from a good meal. Instead of ruining the health of the owner, as cars do, they actually improve it.

It is a myth that cycling can only be popular

POSITIVE ACTION

Making the most of the bicycle

● **Cycle to work**
If you drive to work, ask yourself if you could make the journey by bicycle. Cycling is considerably less stressful and much more invigorating than sitting in a car.

● **Use cars for long distances only**
Think of your cycle as the vehicle for local travel. Cars should only be used if other forms of transportation are inappropriate.

● **Press for cycling facilities**
If you live in a built-up area without cycleways, suggest to your local traffic planners that some are created.

---

where the land is flat. Certainly, if you live in Holland, for example, you can enjoy some of the easiest and most energy-effective travel that there is. But what would happen if planners put as much work into cycleways as they do into roads? Why should cars – which after all cause so many problems – enjoy all the investment?

Constructing permanent cycleways through cities and rural areas is a cheap way of reducing pollution and congestion. With good planning many of the problems cited against bicycles – the risk of accidents, pollution and theft – can all be solved. Why can't employers provide bicycle parking space for employees as they often do for cars? Why can't substantial sections of inner cities be designated "bicycle only" areas? Why can't weatherproof paths be provided for cycling in winter?

We simply cannot afford the ever-greater proliferation of the private motor car. If we don't encourage some imaginative thinking about alternative ways of getting around, we shall end up like the three-quarters of the world who still have only one form of transport: their feet.

Pollution-free travel Cycling is an entirely non-polluting method of travel: it deserves far more attention from planners than it currently gets.

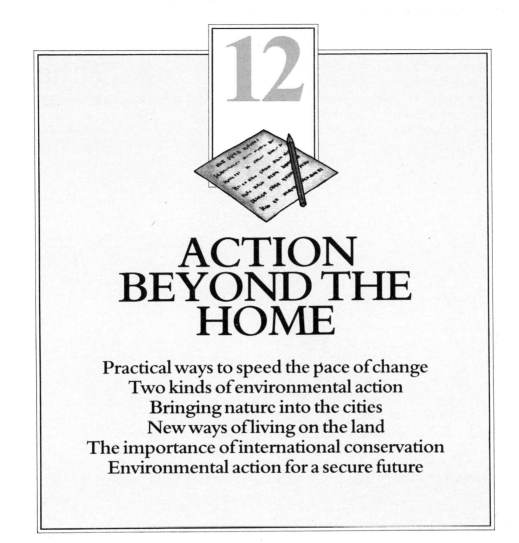

12

ACTION BEYOND THE HOME

Practical ways to speed the pace of change
Two kinds of environmental action
Bringing nature into the cities
New ways of living on the land
The importance of international conservation
Environmental action for a secure future

So far in this book, we have dealt with environmental action in terms of individual choice: how you can improve the environment by making small but very important changes in the way you live. This kind of change is gaining ground. More and more people have seen through what might be described as the pimply adolescent stage of human development and now realize that felicity does not depend on expensive gadgets and knick knacks. An increasing number of people accept that this planet was not just created for the selfish enjoyment of *Homo sapiens*, but that other creatures have a right to live on it as well. We cannot kick the ladder of nature away from under us, for we are part of nature too.

But there are those who will tell you – for whatever reason – that just putting your own house in order is not enough. The changes this will bring about, they say, will be too slow and too limited to rescue our planet from the grave problems that currently confront it.

Well, in one sense they are right. Putting your own house in order, *on its own*, will not solve the problems that the world faces. But doing this, along with thousands or millions of others, and at the same time acting on a broader front, certainly will.

The pace of change can be quickened immeasurably if you reach out beyond your own walls and fences to try to influence what is happening in the rest of the world. Bob Geldof showed what can be achieved when millions of people contribute to an international cause. Band Aid, Live Aid, and dozens of similar events were enormous fund-raising successes. They have provided money which, as we will see later in this chapter, can help to put right some of the ecological damage that has been inflicted on the world.

Whether through your local community, the country at large, or at an international level, there is much that can be done by ordinary members of the public. The only qualification you need is a determination not to be fobbed off.

At present we live in a life-destroying society. We are living on nature's capital instead of its income. In nature, production exactly matches consumption. Everything that grows eventually decomposes; the death and decay of one being becomes the precondition for the life of another. But our industrially based lifestyle, on the other hand, consists of loose ends. Raw materials are

THE BLUEPRINT IN ACTION
Part 1: Sustainable living on the land

The illustration on this, and the following three pages, shows conservation as it might be put into practice. The principal change that would be seen in the countryside is the disappearance of the kind of agriculture shown on pages 40–41. This would be replaced by sustainable organic farming. Instead of producing huge amounts of single crops, even small areas of land would provide a whole variety of farm products.

RECYCLING ORGANIC WASTE
All the manure produced by farm animals would be collected and then taken back to the land where it would be used as a valuable natural fertilizer.

Manure

Free-range livestock

CROP ROTATION
Growing a different crop in each field every year would ensure that the natural fertility of the soil was not depleted and that pests did not have a chance to become a permanent problem.

ORGANIC FARMING
By recycling organic waste and following natural methods of pest control, the need for injurious agricultural chemicals would be avoided.

ANIMALS OUT OF DOORS
Farm animals kept outside would fertilize the soil and use up any organic household waste. They would supply a useful amount of food for a minimum amount of investment.

converted into consumer products, and these are not recycled. Instead they end up as sources of pollution.

Only by making conservation a worldwide issue can we change our society into one which protects and which finally renews the Earth's living resources. We must not only consume less, but we must learn to consume differently, and we must persuade others to do so as well. If this change can be made, the result will not be some Utopia, but at least it will be an Earth which is worth handing down to future generations.

TWO KINDS OF ENVIRONMENTAL ACTION

If the environment shows signs of ill health, there are two ways in which it can be treated. First there is direct action: protecting local wildlife, replacing trees, clearing away garbage, preserving open spaces, and so on. These are emergency operations – measures that tackle acute symptoms. This kind of

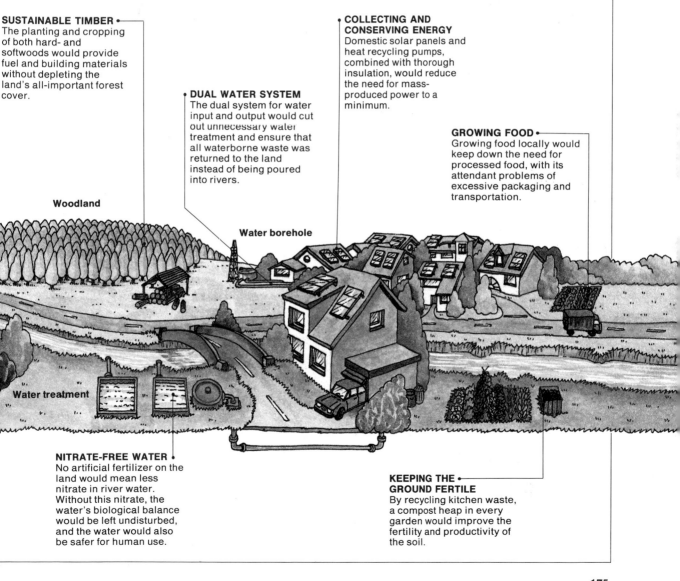

SUSTAINABLE TIMBER
The planting and cropping of both hard- and softwoods would provide fuel and building materials without depleting the land's all-important forest cover.

DUAL WATER SYSTEM
The dual system for water input and output would cut out unnecessary water treatment and ensure that all waterborne waste was returned to the land instead of being poured into rivers.

COLLECTING AND CONSERVING ENERGY
Domestic solar panels and heat recycling pumps, combined with thorough insulation, would reduce the need for mass-produced power to a minimum.

GROWING FOOD
Growing food locally would keep down the need for processed food, with its attendant problems of excessive packaging and transportation.

Woodland

Water borehole

Water treatment

NITRATE-FREE WATER
No artificial fertilizer on the land would mean less nitrate in river water. Without this nitrate, the water's biological balance would be left undisturbed, and the water would also be safer for human use.

KEEPING THE GROUND FERTILE
By recycling kitchen waste, a compost heap in every garden would improve the fertility and productivity of the soil.

action is vitally important, and anyone who seriously wants to improve the world around them can achieve enormously worthwhile results by joining a local conservation trust. Over the last decade or so these have sprung up by the hundred: some useful addresses appear on page 185.

But important though this kind of action is, it is only one-half of the job. The other half is concerned with changing the way in which we use the environment, so that it is actually renewed. This is prevent-

ative medicine, and in the long run, it is the only way to cure the patient. This kind of action may not bear immediate results, but it is an investment which in the future will yield handsome dividends.

IMPROVING YOUR LOCAL ENVIRONMENT

Throughout this book, we have seen how waste and pollution are caused by the folly of the "big-is-better" mentality, by the stretching of supply chains so that producers and consumers are hundreds, if

THE BLUEPRINT IN ACTION
Part 2: Sustainable living in the city

In cities, the principal effects of full-scale conservation would be a great reduction in energy consumption, atmospheric pollution and the production of waste. As much energy as practicable would be provided from renewable sources such as wind and waves. A reduction in the use of cars would improve the atmosphere, while systematic recycling of waste would ensure that far less of it was uselessly thrown away.

THE WATER SUPPLY
Unpolluted river water would require no chlorination and little other treatment before being piped to houses and factories for all uses except drinking.

Recycling plant

RECYCLING WASTE
All recyclable materials – glass, metals and paper – would be collected separately and processed in a recycling plant for re-use.

BACK TO THE LAND
Organic waste from the urban sewage system would be transported back to farmland for use as fertilizer.

NATURE IN THE CITY
Parks and waste ground would be planted with trees. These would play a part in cleaning the urban air and would also provide habitats for wildlife in the city.

not thousands, of miles apart. With all our daily needs – water, energy, but above all, food – we are far removed from the sources of supply. As a result the environment suffers – in the case of food, through mass production, mass haulage, and mass waste. This is where preventative treatment can be applied on a local level – in helping to close the local cycle of production, so that you have a say not only in *what* is produced, but also *how* and *where*.

Living more locally may sound rather abstract or inward-looking, but in reality it is neither of these things. It does not mean shutting yourself away in some remote area and never having any contact with the outside world. It simply means drawing on the resources close at hand first, and then if need be, drawing on those from further afield.

The American Briarpatch movement, which began in the mid-seventies, was an early example of local living in action. By selling local produce, Briarpatch stores brought producers and consumers

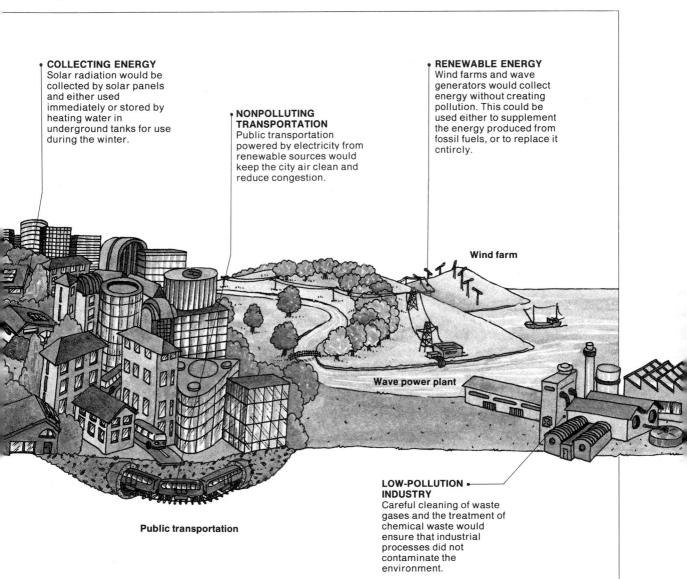

COLLECTING ENERGY
Solar radiation would be collected by solar panels and either used immediately or stored by heating water in underground tanks for use during the winter.

NONPOLLUTING TRANSPORTATION
Public transportation powered by electricity from renewable sources would keep the city air clean and reduce congestion.

RENEWABLE ENERGY
Wind farms and wave generators would collect energy without creating pollution. This could be used either to supplement the energy produced from fossil fuels, or to replace it entirely.

Wind farm

Wave power plant

Public transportation

LOW-POLLUTION INDUSTRY
Careful cleaning of waste gases and the treatment of chemical waste would ensure that industrial processes did not contaminate the environment.

closer together, and so reduced the need for great amounts of packaging, preservatives and fuel. Briarpatch customers knew what they were buying and where it was grown. The Briarpatch movement was ahead of its time, and suffered for it. But since then, the ideas behind the Briarpatch movement have become even more relevant, and the kind of local economy that the movement pioneered now seems certain to gain strength. Local living improves the quality of the land and the quality of its produce. It also provides employment, which in these days of enforced idleness, is surely a good thing.

In Holland the *Kleine Aarde* (Small Earth) movement has done a similar job to the Briarpatch movement, but has gone a little further by encouraging people to set up small businesses, setting up shops in many towns and encouraging the production and consumption of organic produce.

Keeping things local is something that everyone can play a part in. By getting together with a group of like-minded people, you could buy a share in a farm that would supply you with much of your food – food that is not contaminated with chemicals. Starting up such an enterprise only requires a few people to commit their savings and enthusiasm. If enough people do this worldwide, much of the unnecessary and ecologically damaging transportation of goods over huge distances will be prevented.

GREEN CITIES

Of course, taking everyday practical action to improve the state of the countryside is only possible if you live in the country. But in the cities, too, there is a great deal that the individual can do to improve the environment. The long-term survival of our species requires that all of us – whether we live in the country or the city – understand and appreciate the complexity of nature. Now with the best will in the world, you cannot do that if you are surrounded by concrete and asphalt. The practical solution is to help nature to re-establish itself, even a little, in places from which until now it has been banished.

In countries like Britain, which was among the first to experience rapid industrial and urban growth, and, more recently, inner-city decay, conservation groups have turned abandoned land into wildlife parks. At the William Curtis Ecological Park in London's docklands, city people have a chance to see for themselves the wild animals and plants which once lived on the banks of the Thames before they were overwhelmed by roads, warehouses, and office buildings. Similarly, on urban farms in a number of countries, children who may not even have seen a cow before can find out about farm animals and grow plants – all essential if they are to grow up with any understanding of the importance of the natural world.

Few American cities have grown organically like those in the Old World. Most are concentrated accumulations of buildings with few green spaces or even trees. Here, contact with nature may be even harder to find. But in the United States the conservation movement and wildlife groups are trying to redress the balance, so that there is room for nature as well as humans.

If everybody, including city planners, were to put daily contact with the natural world high on their list of priorities, how different city life would be. In "green" cities, trees would be planted to form natural corridors that would link neighborhoods. Forgotten rivers would once more flow on the surface instead of being piped deep below ground, while in their place traffic would flow out of sight.

The ideal of such green cities may be a long way from being realized, but in the meantime, those who live in cities can help to speed up the process of change. As well as actively encouraging wildlife by the proper use of open spaces, there is the battle against the greatest enemy of the greening of the cities – the motor car. With its insatiable demand for space and with the pollution that it causes, the rule of the car must be challenged. Campaigns like "Cities for People," which we encountered in the previous chapter, are a sign that this process is beginning. When enough people demand that the car is used *intelligently*, the development of true green cities will not be far off.

NEW WAYS OF LIVING ON THE LAND

During the first part of this century, there was a tide of emigration from the countryside to the cities. There is evidence that in the developed world this

Nature conservation in action
The William Curtis Ecological Park (*left*) in London gives wildlife a toehold in what is otherwise a purely man-made environment. Wildlife road crossings (*above*) enable animals to survive another of man's intrusions.

tide has turned, and that more and more people are moving out into rural areas. On the face of it, this sounds all well and good. But too often it has meant cramming new arrivals into houses that are cut off from the land around them, like islands in a sea of agribusiness. Furthermore, in most cases their occupants have to travel back into towns and cities to work, so their life is not nearly as rural as it might appear. But it need not be like this, as some people have discovered.

When real rural life was still a common experience, most people had an area of land which they could use for producing food. This would not be their only means of livelihood – they would have had some other means of generating income as well – but they made sure that their land worked for them. In a few experimental villages in America and Europe, some attempts are being made to repeat a highly successful pattern from the past, so that people do not simply rely on paid work.

At Lightmoor on the outskirts of Telford New Town in Britain, a new, self-reliant village is being built. The future inhabitants, with the help

Improving the environment The voluntary work of wildlife trusts and other conservation groups is essential if the countryside is to survive the neglect and overexploitation caused by industrial farming.

of architects and builders, are constructing their own energy-efficient houses. These are designed to include enough workshop or office space so their owners can work at home to earn a living, at least on a part-time basis. The houses also have gardens which are large enough to grow a fair amount of food. Instead of working for all their food, they will be able to grow some of it.

. If you are moving from the city to the country, this way of getting back to the land is a practical method of improving the environment through your own handiwork. It is a halfway house to self-sufficiency, a way of living which demands less input from the merchants of mass production, and which therefore creates less pollution and waste. Growing all your own food is very hard work; growing a worthwhile part of it should not be beyond the capabilities of anyone with a patch of land and a degree of energy.

INTERNATIONAL CONSERVATION

These then are just a few of the ways in which you can help to improve the state of that part of the world which is immediately about you. But what about the next step up – the great jump from local matters to matters that concern the whole country? How can you play a part in environmental action on a national or even international level?

Governments do not, on the whole, like conservationists. And when conservationists are scattered and disorganized, they find it easy enough to deal with them. But when conservationists become organized, the situation changes rapidly. So the best way that individuals can help to influence national policies is by swelling the membership (and funds) of environmental pressure groups. In Europe, there are now more members of such groups than there are members of political parties, and we are beginning to see the results.

Before the growth of organizations such as Greenpeace and Friends of the Earth, any protests about the way the environment was being treated could be quashed by bringing in "experts" who would supply reassuring statistics and prognostications. This still goes on. But establishment experts today have a much more difficult time of it. The reason for this is that large environmental organizations

---

POSITIVE ACTION

What you can do to help international action against pollution

- **Support practical action**
  The most effective step you can take to help is to join one of the major global environmental groups, such as Greenpeace (see p. 185).

- **Complain about product pollution**
  Many companies import products like aluminum, biocides, and plastics which cause considerable pollution in their countries of manufacture. If you have such a company near you, write and ask them if they have considered using less polluting alternatives.

- **Help to stop the international waste trade**
  A number of countries, including Britain, take in toxic waste from other countries for disposal. The transportation of toxic waste is highly dangerous. If you live near a factory that accepts waste from abroad, write and point out that whoever creates toxic waste should have to deal with it *themselves*.

- **Take action abroad**
  If you notice pollution when abroad, don't keep it to yourself – complain! Pollution of the environment is an international problem – something that everyone has a right to object to.

---

now have highly qualified staff of their own who can take on the experts at their own game. Conserving the environment has become a professional business, thanks to the growing number of people who join the conservation organizations.

As well as helping to promote change with membership subscriptions and contributions to funds, you can help by complaining – by writing and objecting to specific activities that are harming the environment. Writing letters may not sound like a very active way of promoting change, but it works. The power of the pen and the power of the purse go hand in hand.

WORLDWIDE CONSERVATION IN ACTION

Organized conservation and anti-pollution action have had some spectacular successes. In 1980 over 200 million people saw a television program which probably saved one of the world's greatest animal species – the humpback whale – from extinction. The film of its behavior and the record-

Direct action against pollution Direct practical action has become an important part of the struggle against pollution. Here members of the Greenpeace organization are seen during an attempt to block the dumping of mining waste on Spain's Mediterranean coast. This single pipe had covered the sea bed with a daily discharge of 7,000 tons of mining waste rich in toxic lead and zinc, and as a result local fish stocks had suffered heavily. Direct action has proved an effective way of drawing the world's attention to pollution black spots and areas where wildlife is endangered.

ing of its song turned out to be one of the most evocative appeals for nature conservation. Friends of the Earth lobbied politicians all over the world to outlaw whaling. Greenpeace took direct action by following whaling ships across the oceans and by launching motorized inflatables in which they positioned themselves between whaling ships and whales. In this way they prevented the slaughter of some of these animals at the time, and helped ensure the prevention of such slaughter in the future.

The sudden change that was brought about in whaling was an example of the power of the purse and pen working to the greatest possible effect. Huge numbers of people subscribed to the campaign. For years politicians had shilly-shallied but then, quite suddenly, they acted. They didn't act because they had all become environmentalists: they acted because they were in danger of being outflanked by millions of people who wouldn't take no for an answer. In the space of a few years nearly all countries banned whaling.

Similarly, international action against pollution has made some progress. The growing campaign against dumping nuclear waste at sea has made many governments think twice about getting rid of radioactive substances in the world's oceans.

THE IMPORTANCE OF SAVING WILDLIFE

We have already reached the point where we are causing the extinction of a species every day, most rapidly in the shrinking tropical forests. By the end of the century it will be one species every hour, if present trends continue. But what, some people may ask, is the point of saving wildlife? Why should we spend our money on saving animals that are of no use to us?

Well, apart from the aesthetic and moral considerations which surely forbid us from removing any animal species from the face of the Earth, there are other far more practical considerations to be borne in mind. Animals are living indicators of the health of the environment. They are our first line of warning when things are going wrong, because they are far more sensitive to environmental changes than we are. When seabirds suddenly abandon their traditional breeding grounds, or when with equal suddenness predatory mammals begin to vanish, man-made pollution and environmental destruction are often to blame.

The fate of the Bengal tiger shows how the fate of wildlife and man are inextricably entwined. Saving tigers from extinction was one of the tasks that the World Wildlife Fund (now the Worldwide Fund for Nature) set itself after it came into being in 1961. By 1973 the Bengal tiger in India had dwindled to a population of less than 2,000 from an estimated 40,000 in 1900. But it was not just the tiger that was threatened. The disappearence of the tiger, which is the predator at the top of the forests' food chains, was a signal that the forests were becoming ecologically unbalanced, and that they were not going to survive. And if the forests went, so too would the firewood and eventually the soil on which all life, including human life, depends.

As part of Project Tiger, fifteen tiger reserves were created, and the forest within them was left undisturbed by the activities of hunters, farmers, and loggers. By 1983, this had restored the tiger population to 4,000 animals. Within the reserves the tiger population almost quadrupled.

Not only the tiger had been helped – the environment as a whole benefited. The World Wildlife Fund recorded some of the changes that had happened in a report published on their 25th anniversary. "Streams that used to flow only during the monsoon became perennial and, compared with those in adjacent forests, were strikingly free from siltation. Vegetation recovered quickly, resulting in the formation of humus and increased fodder for other wild species. Deer, elephant, buffalo, and rhino numbers increased; so did the diversity of species of plants and animals." The project rescued a complete environment.

A CASE OF COMMON INTERESTS

Today the Indian tiger reserves extend to a total of nearly 9,650 square miles. But during Project Tiger, 6,000 villagers had to be moved out of the reserves in order to make their success a real

---

POSITIVE ACTION

What you can do to help international wildlife conservation

- **Join a conservation group**
  The most obvious, but nevertheless most effective, first step in helping international conservation is to join a group (see p. 185).

- **Don't buy wildlife souvenirs**
  Tourism can often have a devastating effect on wildlife through the sale of souvenirs. Buying shells, animal skins, furs, ivory, and corals will all contribute to the destruction of wildlife. Do not buy any of these things and tell other people why they should not buy them either.

- **Don't buy endangered animals**
  When buying any animal from a pet shop, make sure that it has been bred in captivity, and not imported. Reptiles and birds are particularly endangered by dealers who break the regulations on the import and sale of animals from the wild.

- **Don't buy endangered plants**
  The import of garden plants can have a damaging effect on the wild flora. If you are buying rare plants – particularly ones from the tropics – for your garden or home, make sure that they are nursery bred.

---

possibility. At first sight this sounds like a clash of interests between humans and wildlife – one that, most unusually, wildlife was allowed to win. But the outcome has been good for both beast and man.

Creeping environmental destruction can be halted by looking at what we need from the natural world, and then making sure we take it in a *sustainable* way. In Project Tiger, the Indian government certainly paid a financial price to move people out of the reserves. But it also gave assistance in improving farming practices and, perhaps most important, in constructing bio-gas plants. These enabled the villagers to stop relying on firewood for their fuel. So, with a little appropriate technology, the forests have been safeguarded.

In another project, the London-based Fauna and Flora Preservation Society and the World Wildlife Fund worked together to save two rare species of antelope – the scimitar-horned oryx and the addax. In 1978 both these species were on the verge of extinction as a result of the civil war in the Central

African country of Chad. In neighboring Niger, the antelopes were also suffering because their habitat was being destroyed by people in search of firewood and building timber.

As with the Bengal tiger, the rescue plan concerned not just the animals, but the whole environment. Local people were introduced to house designs that made use of mud bricks, so they did not have such a need for building timber. They were encouraged to collect firewood in a sustainable way by harvesting it, instead of by cutting down trees. What started as an effort to save two species of antelope grew into far broader ecological action.

These projects are just two of hundreds of similar ones which have been set up by national and international organizations around the world in an effort to link wildlife protection with sustainable use of the land.

GREEN AID

As we have seen, conservation is of two kinds — emergency and preventative. On its own, emergency action to combat individual environmental problems, however admirable and well-meaning it may be, is not sufficient to reverse the trends of environmental deterioration. Even if we succeed in saving individual species, even if we make serious inroads into the global pollution problem, even if we get to grips with soil erosion, and even if we manage to slow down deforestation, we shall have to undo the damage that has *already* been done. Forests worldwide have shrunk from about 75 percent of the land's surface just after the last Ice Age to about 25 percent today. Much of this land has turned into desert or infertile scrub, the rest produces the food that keeps us alive. Close to five billion people have to be fed today and there may be six billion by the end of the century.

Planetary housekeeping is a skill we have yet to understand properly. Projects which renew the environment by undoing past bad housekeeping are our best hope for the future. Whether in the Third World or the First, the renewal of habitats is essential work.

Each of us, whether we like it or not, spends a great and increasing amount of money every year on our national security. But there is another kind of security which is just as important and that is ecological security. This security comes of living in an environment which is in a stable condition and which is not being poisoned, stripped of its vegetation, or washed away. In the First World, our ecological security is at best flimsy. But in most of the Third World, it is now nonexistent. However, individual financial help, when multiplied thousands or millions of times, can play a part in bringing it back.

Ecological security can be regained by improving the way that the land is used. The Greenbelt movement in Kenya is a perfect example of conservation that is designed to do this. The movement is not just concerned with putting a stop to deforestation and soil erosion, but also with reversing the trend by planting new trees on farms, as shelter belts, and on vulnerable slopes. Greenbelt volunteers have started "agroforestry" projects which combine trees and annual crops in one sustainable system of cultivation without the input of expensive and potentially damaging artificial fertilizers.

Agroforestry is an ancient system of cultivation. But for decades, even centuries, European experts have sought to persuade, and sometimes bully, Third World farmers into adopting more "efficient" farming methods. These invariably involve the clearance of all vegetation so that the land can be ploughed. Nowadays both clearance and ploughing are done with expensive imported tractors, which require expensive imported fuel and equally expensive spare parts.

In Africa, Asia, and Latin America the traditional methods of interplanting trees and annual crops have thus been given up in favor of "bare earth" farming. But there is no doubt that for large parts of the tropics agroforestry is the *only* really viable method of cultivation. Bare soil simply cannot stand being exposed to the extremes of intense heat and tropical rainfall. If, on the other hand, the soil is covered by trees, and if these are interplanted with vegetable crops, the soil's moisture and fertility are both improved.

The Kenyan Greenbelt movement has been an inspiration in many other African countries. But all too often people living on lifeless land are too poor to start and sustain effective projects on their own. Just as planting new forests in temperate regions takes foresight and money, so it does in the tropics. Before the beneficial effects of trees make themselves felt, tropical farmers suffer a drop in production – a drop that to someone without any reserves for lean times is unthinkable.

This is where we – the people of the more affluent countries – can do a great deal to help. The money collected by organizations such as Oxfam, together with the huge sums gathered by appeals such as Live Aid, gives Third World farmers a chance to start again. Grants of food and funds keep hungry people alive, but also allow future ecological disaster to be headed off.

In this book we have concentrated chiefly on the side effects of twentieth-century life as they are felt in the industrialized world. But the greatest ecological disasters are taking place in the Third World: renewing the environment in these stricken areas will help not only the people directly affected, but all the world's inhabitants.

TOWARDS A GREEN FUTURE

Making a really lasting improvement in the state of the world is not a pocket-money business. It is something that will require all the resources that can be devoted to it. All our efforts will have to go into revitalizing our ravaged environment by cutting out pollution and waste and bringing back at least part of the Earth's natural clothing of vegetation. This will not only reestablish sustainable conditions for human life, but will also create safe habitats for our fellow species once again.

But initiatives on this scale will require the determined efforts of large numbers of people. In the meantime our immediate task is to do everything we can to bring our present life-destroying culture to a halt. Time is running out dreadfully fast. We have short memories: by the time this book comes out, the horror of Chernobyl will have joined the horror of Bhopal in the dim recesses of our recollections, no longer to affect us or our actions. Perhaps a fresh horror will have taken their places, and become a nine-day wonder – who knows? There will be a Chernobyl in the West sometime – now or in a hundred years – the mathematics of chance make it certain that it will come. Will *that* be enough to make us pause?

We, all of us, have it in our hands to destroy, to protect, or even to renew the Earth's living resources. Through our daily way of life, we help to determine whether we are going to leave our children a devastated planet, or one which is a worthy home for all living creatures. Let us make no mistake about it: the continuation of our present way of life will have dire consequences for all of nature, including our offspring. The combined impact of deforestation, desertification, the acidification, contamination, and erosion of the soil, the pollution of watercourses, and the destruction of living species adds up to a massive crime against posterity. We have invented an eighth deadly sin: *crime against the future*. Never before has it been possible for one generation to do so much damage in such a profound and permanent way.

The present time – just these few decades in which we have been fated to live our lives – is the most crucial time ever for life on this planet. We hold the future of all life in our hands, indeed, we must decide if there *shall* be a future. This must be the consideration that shapes our views and helps to determine our actions. The process of ecological renewal is just beginning: it is up to every one of us to make sure that it does not fail.

USEFUL ADDRESSES

It is impossible in a book of this size to list all the societies and organizations concerned with protecting and improving the environment, especially as their number is increasing all the time. This list includes a selection of national organizations; many of these have active local groups, and all welcome financial support and participation from new members.

American Farmland Trust
1717 Massachusetts Ave., NW, Suite 601, Washington, D.C. 20036.
Informs the public about soil erosion and other threats to the country's agricultural land; cooperates with other groups to preserve farmland.

The Center for Science in the Public Interest
1755 S. Street, NW, Washington, D.C. 20009.
Scientists and lawyers at the Center investigate food, nutrition, and other consumer issues, publishing reports and filing lawsuits as necessary. Its newsletter is *Nutrition Action*.

The Cornucopia Project
33 E. Minor Street, Emmaus, PA 18049.
A research project examining how to develop a more natural and sustainable food supply; promotes organic farming and locally grown food.

The Cousteau Society
930 W. 21st, Norfolk, VA 23517.
Promotes the preservation of the world's oceans and their marine inhabitants. The Society produces films, books, and articles and funds research expeditions.

Defenders of Wildlife
1244 19th Street, NW, Washington, D.C. 20036.
Founded to protect the country's wildlife and wildlife habitat. Publishes *Defenders* magazine.

Environmental Defense Fund
444 Park Avenue South, New York, NY 10016.
This national group's lawyers, scientists, and economists work on all manner of environmental and public health problems, from air quality to toxic chemicals.

Farm Animal Reform Movement
P.O. Box 70123, Washington, D.C. 20088.
Works to stop abuses to farm animals and other ill effects of modern agriculture.

Greenpeace U.S.A.
2007 R Street, NW, Washington, D.C. 20009.
Initiates direct action campaigns to protect the marine environment and marine animals from pollution and decimation.

Institute for Local Self Reliance
1717 18th Street, NW, Washington, D.C. 20009.
This research and consulting group promotes projects, like recycling, that build neighborhood self reliance.

National Audubon Society
950 Third Avenue, New York, NY 10022.
Promotes wildlife preservation through research, education, and action programs. Maintains 76 wildlife sanctuaries, many nature and research centers, and camps. There are 500 local chapters; publishes *Audubon* magazine.

National Trust for Historic Preservation
1785 Massachusetts Avenue, NW, Washington, D.C. 20036.
Uses advocacy, education, financial aid, and technical assistance to preserve historic buildings and neighborhoods.

Natural Resources Defense Council
122 E. 42nd Street, New York, N.Y. 10168.
NRDC's lawyers and scientists keep track of government agencies, file lawsuits, and publish information on a variety of environmental problems. Its magazine is *Amicus Journal*.

The Nature Conservancy
1800 N. Kent Street, Suite 800, Arlington, VA 22209.
Acquires natural lands to preserve biological diversity; works with the states to identify ecologically significant areas. Manages 800 nature sanctuaries.

Oceanic Society
Stamford Marine Center, Magee Avenue, Stamford, CT 06902.
Conducts education and research to further protection of the marine environment. Publishes *Oceans* magazine.

Sierra Club
730 Polk Street, San Francisco, CA 94109.
Club volunteers in 55 chapters and hundreds of groups work on the gamut of environmental problems, assisted by a national staff of lobbyists and organizers. It publishes books and *Sierra* magazine and runs an outings program.

The Trust for Public Land
82 Second Street, San Francisco, CA 94105.
This private group buys open land in cities and in the countryside for eventual transfer to public ownership. It also helps local groups set up community and agricultural land trusts.

Worldwatch Institute
1776 Massachusetts Avenue, Washington, D.C. 20036.
This research organization publishes the *Worldwatch Papers*, a series of reports which identify and analyze global problems and trends.

World Wildlife Fund—U.S.
1600 Connecticut Avenue, NW, Washington, D.C. 20009.
Concentrates on international wildlife conservation; associated with the World Wildlife Fund International.

FURTHER READING

In this list of suggested further reading, we have included books from as wide a range as possible, some old, some new, some general, some specific. There are classic works such as E. F. Schumacher's *Small is Beautiful*, together with more recent publications. The books are arranged under general subject headings and we hope that they will provide something of interest to everyone.

ECOLOGY AND THE ENVIRONMENT

Brown, Lester *et al*, *State of the World, 1986*; A Worldwatch Institute Report; W. W. Norton, New York and London

Carson, Rachael, *Silent Spring*; Houghton Mifflin, Boston, 1962

Caufield, Catherine, *In the Rainforests*; Alfred Knopf, New York, 1985

Commoner, Barry, *The Closing Circle* (Nature, Man, and Technology); Alfred Knopf, New York, 1971

Council on Environmental Quality and the Department of State, *Global 2000 Report to the President* (Entering the 21st Century); U.S. Government Printing Office, Washington, D.C., 1980

Ehrlich, Paul R. and Anne H., and Holdren, John P., *Ecoscience* (Population, Resources, and the Environment); W. H. Freeman, San Francisco, 1977

Hyams, Edward, *Soil and Civilization*; Harper & Row, New York, 1976

Kozlovsky, Daniel, editor, *An Ecological and Evolutionary Ethic*; Prentice Hall, New York, 1974

Leopold, Aldo, *A Sand County Almanac*; Oxford University Press, New York, 1966

Leopold, Luna B., and Davis, Kenneth S., *Water*; Time-Life Books, New York, 1970

Myers, Norman, *The Sinking Ark*; Pergamon Press, Elmsford, NY, 1979

Odum, Eugene P., *Fundamentals of Ecology*; Holt, Rinehart and Winston, New York, 1971

Schumacher, E. F., *Small is Beautiful*; Harper & Row, New York, 1973

Thoreau, Henry David, *Walden and Other Writings*; Random House, New York, 1937

ENERGY

Anderson, Bruce, and Riordon, Michael, *The Solar Home Book*; Brick House, Andover, MA, 1976

Clark, Wilson, *Energy for Survival*; Anchor Press/Doubleday, New York, 1974

Lovins, Amory, *Soft Energy Paths* (Toward a Durable Peace); Harper & Row, New York, 1979

Rothchild, John, *Stop Burning Your Money* (The Intelligent Homeowner's Guide to Household Energy Savings); Random House, New York, 1981

FOOD AND NUTRITION

Brody, Jane E., *Jane Brody's Good Food Book*; W. W. Norton, New York and London, 1985

Kirschmann, John, D., with Dunne, Lavon J., *Nutrition Almanac*; McGraw-Hill, New York, 1984

Schell, Orville, *Modern Meat*; Random House, New York, 1984

van den Bosch, Robert, *The Pesticide Conspiracy*; Doubleday, Garden City, NY, 1978

Winter, Ruth, *A Consumer's Dictionary of Food Additives*; Crown, New York, 1984

HEALTH

Brody, Jane E., *The New York Times Guide to Personal Health*; Times Books, New York, 1982

Editors of Prevention magazine, *Fighting Disease*; Rodale Press, Emmaus, PA, 1984

Inglis, Brian, and West, Ruth, *The Alternative Health Guide*; Alfred Knopf, New York, 1983

Samuels, M.D., Mike, and Bennett, Hal Zina, *Well Body, Well Earth*; Sierra Club, San Francisco, 1983

Vickery, M.D., Donald M., and Fries, M.D., James F., *Take Care of Yourself*; Addison-Wesley, Reading, MA, 1981

LIFE ON THE LAND

Berry, Wendell, *The Unsettling of America* (Culture and Agriculture); Sierra Club, San Francisco, 1977

Goldstein, Jerome, ed., *The Least is Best Pesticide Strategy*; The J. G. Press, Emmaus, PA, 1978

Merrill, Richard, ed., *Radical Agriculture*; Harper & Row, New York, 1976

Staff of Organic Gardening magazine, *The Encyclopedia of Organic Gardening*; Rodale Press, Emmaus, PA, 1978

INDEX

plastic, *80–1*, *83*, 86–9
recycling, *83*, 84–5, 87–8, 89–90, *91*, *176*
gardening, 127–38, *128–38*
  organic, 135–8, *135–7*
  pesticides, *14–15*
gas, natural, 28–9, 30, 48, 49, *140*, *147*, 168–9
gas stoves, *118*
gasoline, 87, 92
  lead content, 104, 161–2
  see also oil
Georgswerder, 92, 120
geothermal energy, 156
Gerard's *Herbal*, 111
Germany, 100, 118
  acid rain, 11, 143, 145, 146
  biocide residues in food, 70
  building materials, 122
  car accidents, 161
  car exhausts, 160
  garbage disposal, 84
  industrial waste, 92, 120
  lead-free petrol, 162
  nuclear waste, 148
  pharmaceutical industry, 111
  solar energy, 152
  water pollution, 34
  wind power, 155
germs, 93–5, 98
Gerson, Dr. Max, 110–11
glass: recycling, *80–1*, *83*, 89, *91*
  waste disposal, *88*
gluc, 119
glue-sniffing, 106
Goodfellow, F., 145–6
governments, 13
"green" cities, 178
green manuring, 137
Greenbelt movement, 183
greenhouse effect, 169
greenhouses, *130*
Greenland, 37, 162
Greenpeace, 38, 180, 181, *181*
growth hormones, 52
guano, 28, 48
Guatemala, 74
gypsum, 146

H

hair care, 100
Hamburg, 120
hamburgers, 71–3, *72*
Harvard University, 117–18
Harz mountains, 143
health, 103–12
heart disease, 70, 103, 106, 112
heat recycling pumps, *153*, *155*

heating: domestic, 150–6, *150–1*, *153*
  energy requirements, *143*
heavy metals, 29, 35–6, 84, 92
Helminthosporium maydis, 43
herbal remedies, 109, 110, *110*, 111–12, *112*
herbal teas, *106*
herbicides, *45*, 127–8, 135–6
hexachlorophane, 99–100
Heyerdahl, Thor, 37
Hills, Lawrence, 133
holistic medicine, 109–12
Holland, 120, 145, 152, 168, 172, 178
Honduras, 74
Hooker Chemical Company, 36
hormones, factory farming, 52
household garbage *see* garbage
houses: building materials, 113–17, *114–15*, 122–6, *123*
  chemical hazards, *116–18*, 117–22
  energy consumption, 139–40, *143*, 149–56, *149–51*, *153*, *155*
Humber River, 38
humus, 42, 74
hunter-gatherers, 105
Hyde Park, 36
hydrocarbons, 119
hydroelectric power, *141*, *147*, 152, 156
hydrogen, as a fuel, 170

I J K

Illinois, 48–9
illness, 43, 65–6, 103, 106–7
incineration, garbage, 76, 79–80, *82*, 92
India, 182
Indian Medical Service, 107
Indonesia, 125
industry: energy consumption, *142*
  waste, 35–8, 77, 92
insect pests, 46, 48
insecticides *see* pesticides
insulation, *116*, 122, *123*, *143*, 149–50, *150–1*, *153*
Integrated Pest Control (IPC), 48
interplanting, pest control, 132
Inuit, 52
Ireland, 13
Irish Sea, 148
irradiated food, 66, 67
Israel, 152
Italy, 70, 80
Ivry sur Seine, *82*
Japan, 66, 74, 120, 125
Japan, Sea of, 37
Johns Manville Corporation, 114
Jutland, 155
Kalahari Desert, 105
Kenya, 70, 183

Kleine Aarde movement, 178

L

lakes, acid rain, 11
Latin America, 183
laundry detergents, *94*, 95, *97*, 98 *101*
laxatives, *108*
lead, 67, 92
  in gasoline, 161–2
  in paint, 121, *121*
  in sewage, 29
  in traffic fumes, 104, *159*
  water pollution, 35, 37
"lean-burn" engines, *165*, 166, 168
Lebow, Victor, 77
leguminous plants, 48
leukaemia, 112
life expectancy, 103
lighting, energy requirements, *143*
Lightmoor, 179–80
lindane, 119–20
Lisbon, 110
Lister, Joseph, 93
Live Aid, 174, 184
living standards, 14–16
localism, 16, 177–8
London, 104, 110, 169, 171, 178, *179*
Los Angeles, 23, 104
Love Canal, 36, 92
Lund, 156
lymphocytic leukaemia, 112

M

McCarrison, Sir Robert, 107
McDonald's, 72
macrobiotics, 110
Madrid, 110
Magendie, François, 61
malaria, 110
Malaysia, 125
Manchester, 146
manure, 54–5, *135*
massage, 110
meat, *15*, 52–3, *56*, *59*, *64*, *69*, 71–4, *72*
medicine, 103–12
  see also drugs
Mediterranean, 37, *181*
Merck, 120
mercury, 35–6
Mersey, River, 38
metal: heavy metal pollution, 29, 35–6, 84, 92
  recycling, *83*, 89–90, *91*
methane digesters, 30
methanol, 168, 169
milk, *64*, 70
Mississippi River, 40

ACKNOWLEDGMENTS

AUTHORS' ACKNOWLEDGMENTS
We would like to thank all those authors whose writings we have drawn on, and the many organizations committed to environmental change who were so generous with their knowledge, support and time. We would also like to thank everyone who has helped us more informally with advice and information. We particularly wish to thank Alfred Winter, Kay Morgan, Dr Charles Horth, Dr Lawrence Plaskett, Dr Bernhard Raninger, Dr Chris Gossp, José Lutzenberger and Richard St George. Finally we wish to thank David Burnie, Becky Abrams, Jane Owen and Ian Penney, and through them our publishers Dorling Kindersley, who, with us, were determined to make this an informative as well as a beautiful book.

Dorling Kindersley would like to thank: Melanie Miller and Tim Lang from The London Food Commission for all their help; Brian Price for his scientific expertise; Hester Abrams for her translating skills; Dr Saad Abdalla for medical advice; James McDonald for nutritional advice; Deirdre McQuillan from *Here's Health*; Laura Thomas from CLEAR; Arnold Bell and *Country Living*; Keith Nightingale from SMACMA; Sylvia Beamish from the Anti-Steroid Action Group; Fred and Cathy Gill for proofreading; Hilary Bird for the index and Chris Cope for photographic services.